Building A MINI STOCK Step-By-Step

Editor: Steve Smith

Associate Editor: Georgiann Smith

Cover Photo by: Skip O'Brien, Circle Bee Photography

ISBN 0-936834-52-8

Published by

STEVE SMITH AUTOSPORTS PUBLICATIONS

P.O. Box 11631/Santa Ana, CA 92711/(714) 639-7681

About The Author

Jean Louis Genibrel, a native of France, moved to the United States in 1963.

His racing career started in 1968 with an Austin Healey Sprite which he raced in slaloms and SCCA races.

Since then Jean has driven and built several record holding and championship cars and karts. His successes cover NASCAR, SCCA, IKF, and as a mechanic in IMSA, USAC and SCCA competition.

Genibrel also worked for Dan Gurney's All American Racers during the Bobby Unser-era as a mechanic, and he is currently racing a Mini Stock on the West Coast.

Jean is a consultant and lives in La Verne, California, with his wife Laura and daughters Natalie, Shelley, Stacey and Brandie.

Genibrel has authored two other books for Steve Smith Autosports — The Complete Karting Guide and Racing The Yamaha KT-100S Engine.

Table of Contents

A Special Thanks

I would like to extend some very special thanks to all the people who participated in this project. Most especially are those who have given more than expected of themselves, taking time from their work and families to assist someone else achieve a somewhat selfish goal. Building a race car is in itself, for the average person, a monumental job.

Relating that experience to others in printed form involves thinking for others as well as oneself and this opens the gates to a million and one questions, questions often demanding, often seemingly semi-nonsensical. But, people like Gary Ebeling, and Dwayne and Dan Esslinger took it all in stride, with patience and understanding. Gary, you are more than a great chassis builder, Dwayne and Dan you are more than great engine builders—you are great people.

Space restricts us from printing the names of people and companies who have helped with this project. But some people deserve that extra thank you for their extra effort:

Gary Ebeling (Ebeling Engineering), Dwayne, Dan and Beth Esslinger (Esslinger Engineering), Bryce Russell (Ford SVO, Motorsports), Frank Deiny and Ken Sapper (Speedway Engineering), John Knepp (Electramotive), J.L. Henderson (Carroll Shelby Enterprises), Dick Anderson (Carrera Shocks), Ben Peterson (Peterson Automotive Products), Paul Baker (Baker Precision & Neo Synthetic Lubricants), John Zufelt and Warren Gilliland (JFZ Brakes), Terry Maresh (Modine Radiators), Jack McCoy (McCoy Racing, McCreary Tires), Phil Elliot (Moroso), Jeff Gaul (MSD Ignition), Bob Florine (ARP Fasteners), Dave Dach (ATL), Ken Moore (Colley Ford), Tom Brunn (Ebeling Engineering), John Concialdi (Redline), Steve Mignone (Pyrotect), Vic Irvan (Vic Irvan Enterprises), Bert Robidoux (Usinage Berthold), David Noel and Barry Zacherson (Dana-Spicer), Steve Smith and Geogiann Smith (Steve Smith Autosports), Gary Mathews (Berge, Mathews), Chase Knight (Crane Cams), Terry Davis (TRW Automotive Group), Ron Sammons (Auto Signs Unlimited), Terry Wilson (Tubes Inc.), Mike Spagnola (BAP/GEON), Tex Powell (Tex Racing Ent.), Pennzoil Corp., Roland Smith, Cal Smith.

Thanks
Jean Genibrel

Dedication

This book is dedicated to Cal Smith for his help and special efforts in completing our project car. It is much appreciated.

Jean Genibrel

Introduction

In the late sixties and early seventies, a group of small sedan road racers decided to take to the short circle tracks with their cars. Soon organizations were formed in the South and on the West coast, racing on dirt and asphalt tracks for those four cylinder cars. The most common makes in those days were Austin Mini Coopers, Renaults, Volkswagens, Volvos, Fiats and some Toyotas and English Fords.

When the seventies rolled around with inflation, higher gasoline prices and even widespread shortages, the Mini Stock class looked more and more appealing. Engine and chassis builders brought years of road racing and European rallying experience to the American short tracks and the high banks of Daytona with the "Baby Grand" class (now called the Daytona Dash Series).

The lonely local associations soon grew into a national sport with some tracks featuring the Minis every weekend for the whole season. From the semi stock, quickly thrown together racer emerged the sophisticated tube chassis cars of today.

Chassis builders, like Gary Ebeling, finding the competition intense in the full size modified classes, soon found a niche with Mini Stocks, and with great success. Oval track racing and road racing cars have become more and more sophisticated to the point that the technology in both forms of racing has culminated on a common peak. The cars have also become very similar, to the point that now a GT3 SCCA car can be raced competitively in Mini Stock and vice versa.

Only a few rules separate the two classes but the cars can be changed over to run either form of racing and be competitive. This book was written around the car that was built for this project. While our car was built for oval track racing, it is possible for someone to build a car using this book to run both oval track and road racing.

The differences between the two forms of racing are with the following items: 1) SCCA requires a stock windshield and pillars at the stock angle; 2) minimum weights in Mini Stock racing are determined on a pound per cubic centimeters, usually 1 pound per cc. SCCA determines the minimum weight requirement based on the model that is raced;

3) the SCCA requires the engine to have its number one spark plug in line with the ball joints and most Mini Stock associations allow from 0 to 12 inches of setback; 4) Transmissions are free in Mini Stock, while the GT3 rules only allow 4 speed gearboxes; 5) Mini Stock associations generally allow up to 10 inch wheels and tires, while the SCCA limits wheel width to 7 inches.

Chapter 1

Safety

Whether building a Mini Stocker, SCCA or IMSA car, or any other form of racing car, safety should always be the primary concern. Several companies spend their whole existence preaching safety to their customers. Countless articles and chapters have been written on the subject, and anyone who has ever attended a motorsport event has been acquainted with the dangers of auto racing. It would also appear that the experienced racer would be the most responsive to the needs of safety equipment and designs in his race car — but apparently not so.

While most racing organizations have some rules relating to safety, tech inspections are often lax or even nonexistant. Terms such as "approved" still appear in rule books when relating to fuel cells, driving suits or safety harnesses. Other organizations like the SCCA and IMSA have stricter rules and in-depth safety inspections. The ultimate responsibility lies with the driver.

Safety equipment is designed and manufactured to prevent injuries to the driver in the event of an accident. In this chapter we would like to encompass crash prevention and crash worthiness in the scope of our discussion.

Crash prevention means designing, welding, wiring, plumbing, preparing and testing a car to make it handle, accelerate and slow down to the best of its capabilities to prevent the accident from happening in the first place. The chassis should be set up for optimum handling before the car is raced. After all, if a car is so fast that the driver does not need to drive it to its limits to win, the car and driver will not be near the ragged edge and accidents will not occur as often. And if one should happen, it will not be as severe. Crash worthiness is the ability of a car to absorb a shock created by a wreck without injuring its driver.

Deformability is one of the qualities a racing car should have, while protection of the driver with roll cage, "NASCAR style" side bars, and thoughtful design of the interior are other items that should be taken into consideration while designing a racing car. The redesign of the footwell on Indy Cars in recent years is a typical example of how intelligent design can improve safety.

The safety system falls in five categories: Accident prevention, restraint, heat sheilding, shock prevention, and escape.

1) ACCIDENT PREVENTION This means a racing car is thoroughly tested and prepared so when it hits the track the first time, it will not also hit the wall. The chassis should be set up for optimum handling before the car is raced. Track testing is probably the most practical method.

Some race tracks such as Willow Springs Raceway in Southern California are available on a group or individual basis for testing purposes. There the driver can familiarize himself with the car, its handling and controls, while the crew can sort out the little things that may need attention. A skid pad, such as the one at Willow Springs Raceway, is invaluable in setting up the chassis and learning the limits of the car in a safe environment.

Preventing accidents also means utilizing good welding methods and techniques, utilizing the proper size material, and correct grade of metal. Joining, fastening, and even painting play an important role in keeping a race car together. On our project Mustang, we utilized RBW's L9 series fasteners from ARP. The L9 series of fasteners is still ductile but of greater strength than the grade 8 fasteners. Wheel studs and flywheel bolts of the highest quality, such as ARP's, should also be used.

Taking care that no parts of the suspension or steering bind throughout the complete range of the component will also prevent problems. Good wiring techniques are also important. A loose wire can stall an engine at a critical moment or, worse yet, start a fuel or electrical fire. Very dirty or rusted parts can mask cracks while the rust can weaken the metal. Improperly designed or machined parts will not fit properly and could break at a very inopportune time.

Using the wrong size, type or grade of fastener will lead to trouble. Magnafluxing or other forms of non-destructive testing should be performed on all critical parts of the steering and suspension. Firewalls should be as their name implies — fire proof. And don't leave holes in the firewall for fire to leak through.

The driver himself is often the most dangerous element on the track — both to himself and to others. Burning the midnight oil and expecting to drive at your best the next day is not very sensible. When tired, the mind reacts more slowly and judgement is impaired.

Both are problems that are not what is needed on a race track. The same goes for an angry driver. Short track racing, and even road racing, at times can be a contact sport — and some guys like the contact more than the sport. Being the "contacted" can be infuriating, but driving to get even with another driver can only distract and lead to trouble.

2) RESTRAINT Restraining the driver's body will prevent injuries with the interior of the car, or even a wall, other car or its parts, and at an extreme, the track surface. The restraint system consists of the lap belts, shoulder harnesses, the crotch belt, neck brace, helmet restraint strap, arm restraint and the window net. The restraint system also helps prevent injuries from the inertia generated by a flip or crash.

3) HEAT SHIELDING The driver's fire suit is the most important item. Fireproof underwear, gloves and shoes are also important in preventing painful burns, but in the event of a fire the driver will want to escape promptly. The escape would be very difficult if the driver's hands and feet are burnt.

While a good firewall and sheet metal barrier will prevent the fire from entering the driver's compartment, a good fuel cell (not just an "approved" model like many rule books require), such as an FBI tank and bladder, will take quite a shock before bursting. On-board fire suppression systems such as the one used on our project car from Phoenix Fire Systems will put out a fire in a hurry, or at least allow the driver ample time to abandon the car.

Preventing a fire is still the best way to fight a fire. Do not make the mistake many builders make by using cheap fuel lines and hose clamps while following the rules by using a good fuel cell. Remember that a chain is only as strong as its weakest link. Also, purchase your fuel cell from a reputable company, such as FBI, and install it according to manufacturer recommendations.

4) SHOCK PREVENTION The best way to prevent making contact with the interior of the car is to have the chassis builder take your height into consideration. Allow a minimum of four inches between the top of the roll bar and the top of your helmet.

The rest of the roll cage should be located a far as possible from the driver's head and arms. Remember that under severe impact the neck and spine can stretch as much as 10 percent, and a belt can stretch as much as an inch per foot. All that combined can cause the driver to reach out of the window or over to the passenger side. Or, worse yet, the driver can reach over the roll cage and contact the track surface if the car is upside down.

Contact with the interior of the car can be cushioned with a high density foam, such as the Longacre 6502, to improve shock absorbtion and the foam padding will make entry and exit of the driver's compartment more comfortable.

A window net will help keep the driver's arms and head in the car should his body stretch that far, and the net will also prevent debris from hitting the driver.

The roll cage in this car is well constructed. Note the bar behind the seat to anchor the shoulder harnesses. That bar must have been an afterthought from the looks of the bend in it.

Formula and Indy cars have long demonstrated the value of a deformable structure. Stock car and road racing sedan drivers can benefit from this lesson by building their chassis to bend and absorb the shock of an accident instead of resisting the impact and transmitting the forces to the driver.

Injury can occur even if the driver has not made contact with anything inside the car. As much as 20 G's can be generated for very short periods during impact and the result of such an impact can cause the neck or the spine to stretch as much as 10 percent. These forces are better absorbed by the race car body and chassis rather than the driver.

This chassis is well triangulated and it offers good driver protection. A well designed roll cage built inside a unibody can be as strong as a tube chassis, but it will be heavier and more difficult to repair after an accident. The diagonal tube going into the dashboard could use some padding.

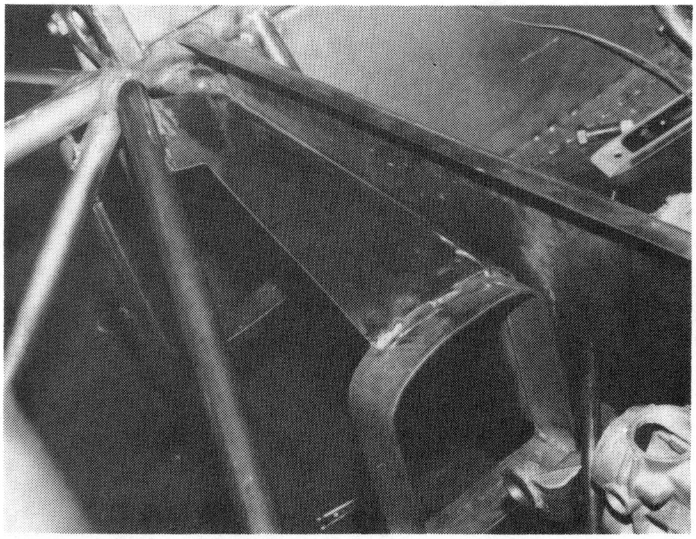

This driveshaft hoop is strong and well designed. The driveshaft tunnel is a good idea.

5) ESCAPE In the event of a fire, the driver will need to leave the car as quickly as possible. The driver will have to hit the fire bottle knob, undo his harnesses, remove the window net, and release the steering wheel. In road racing, some cars are equipped with a two-way radio and a life support system to provide the driver with a few minutes of fresh air in case of a fire (these breathers are available from Competition Systems in Seattle, Wash.). These things need to be disconnected to make a quick exit.

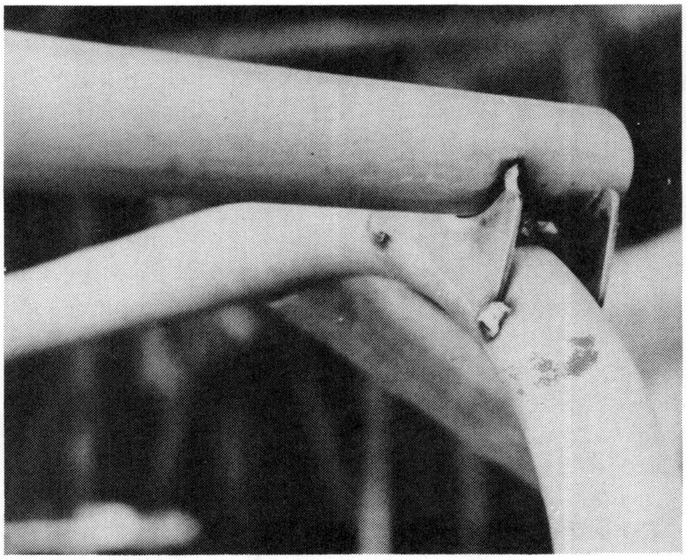

This method of mounting the front roll cage hoop to the roll bar makes for a very strong joint and allows some extra room to the driver's head.

Escaping a crash can be time consuming. The safety minded driver should practice a logical sequence of escape, and store the information in the back of his mind. Frequent mental imagery of this procedure will program the driver's mind to react to an emergency, so even under the most extreme emergency the driver will be able to react without even thinking about what needs to be done.

Seat and Mounts

The racing seat should be mounted to the chassis via a seat hoop. The seat hoop should be mounted to the left on the frame and at the rear of the chassis behind the seat itself. The seat hoop should also be welded in a third location near the right side to prevent pulling up in case of an accident. Such a design will cause the seat to move with the chassis in the event of a strong hit from the side.

The seat should provide support to the thighs, hips, torso and shoulders. The oval track seats that have been used in stock cars for so many years are not as safe as the road racing seats. The ubiquitous stock car seat provides support under the arm pits, against the ribs. In a heavy impact side load the ribs will be heavily bruised, or worse, broken.

The road racing seat spreads the load over the lower part of the body, torso and the shoulders. The human shoulder is much better equipped to take side loads than the rib area is. (Just imagine running into a wall shoulder first. Now imagine running into the same wall with an arm up in the air — Ouch.) The racing seat should also be somewhat flexible but it should not break under a heavy load.

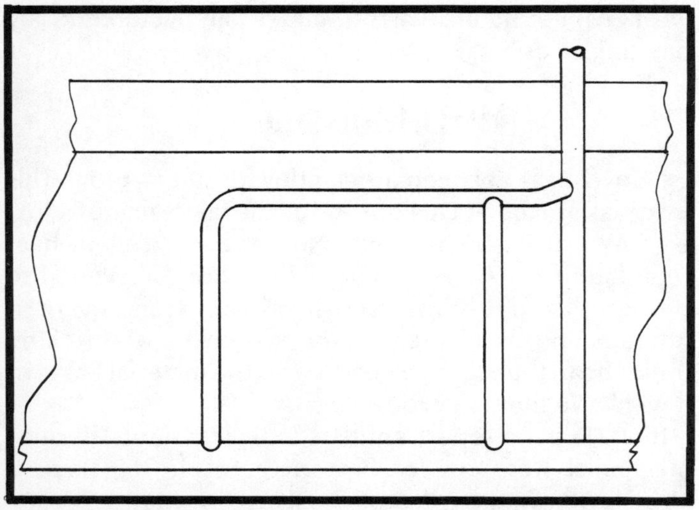

Drawing shows seat mounting frame in top view. The seat frame attaches to the bottom door bar at its left and the roll cage at the rear, and not to the floor of the car. This way the seat and driver will move to the right if another car hits the left door.

Remember that the basic function of the seat is to maintain the driver in a certain predetermined location during a race and in case of an accident, and it should provide driver comfort. Should the seat break up, the driver will no longer be held in place by the safety harnesses.

Aluminum is not as desirable as good fiberglass in seats. The reason is that aluminum will crack and could cut the driver, while fiberglass will bend and absorb some of the shock. No matter what material the seat is made of, it should be enclosed in a strong steel frame.

A seat, simply bolted to the hoop, will tear right out of the mounting bolts under a severe impact. The steel tube frame should provide support to the seat and relieve the stress from the mounting surfaces.

The seat can also incorporate a very important piece of safety equipment — the headrest. It should be strong enough to resist a severe impact by the driver's head. A broken headrest is as good as no headrest at all.

We found a good economy seat at Genesis Fiberglass in Fontana, Ca. Recaro Racing Seats make several models of racing and high performance seats of the highest quality but at a greater cost.

The Headrest

A good headrest, in conjunction with a neck brace, will help prevent injuries to the neck and head in case of an accident. A good headrest should be curved to conform to the curvature of the neck, especially at the base of the skull. In a rear end collision, or if the car should hit the wall backing up, a perfectly flat headrest will come in contact with the rear of the helmet first, pushing the head forward and the neck back. A well designed headrest should provide support to the neck and the rear of the head to prevent the whiplash effect.

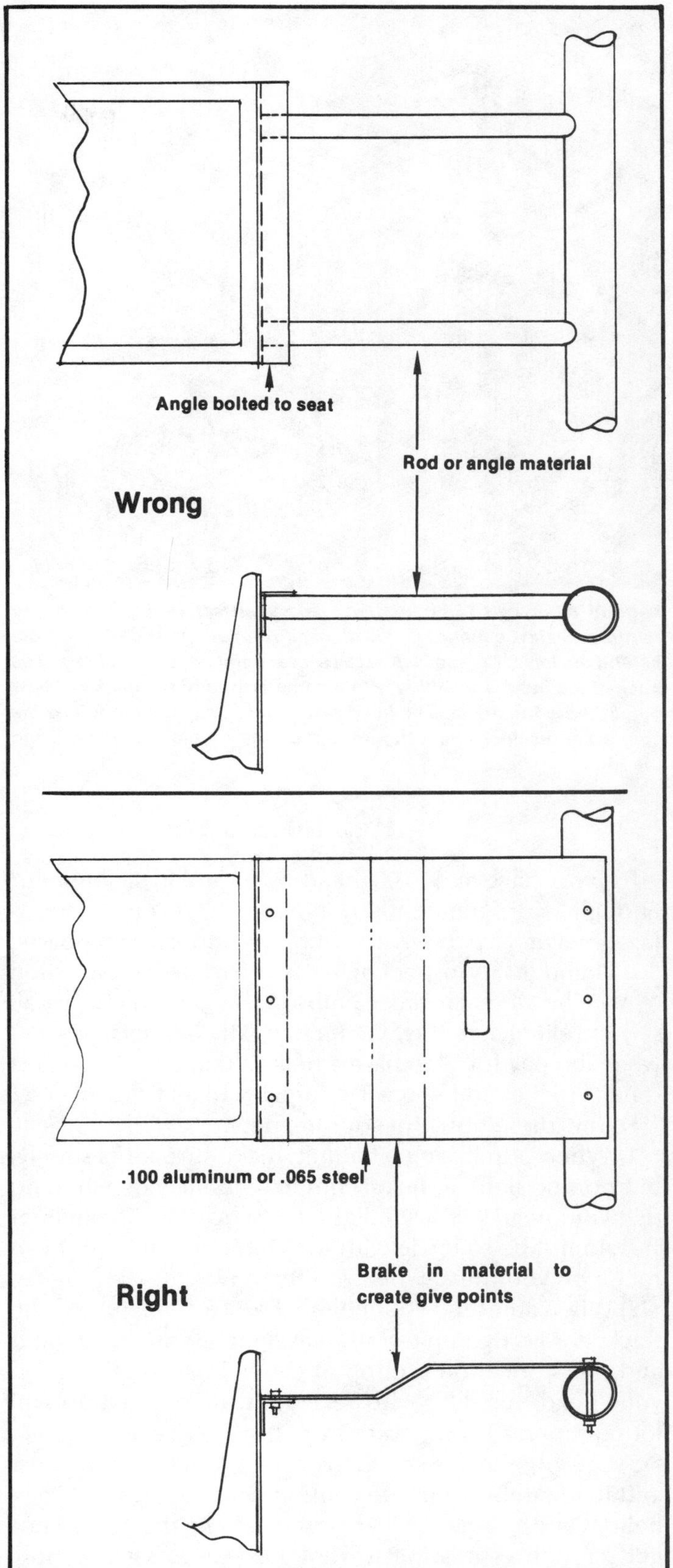

Above are illustrated the right and wrong ways to mount the upper support of the seat to the roll cage. The right way allows for a flexing of the support in case of a hard impact.

A good head rest is imperative. This headrest is made of heavy reinforced plastic material. It is plenty strong but if the car is hit from behind the top of the seat will act as a lever point on the shoulders. The back of the head will collide with the head rest and the neck will bow out, injuring the driver. The head rest should be curved to follow the contour of the neck and it should support the shoulders, the neck and the top of the head

Helmets

"If you have a $10.00 head wear a $10.00 helmet." Enough said? Apparently not. Most short track associations never check drivers' helmets and safety apparel. SCCA and IMSA inspect helmets at each tech inspection — which is at each race. Simpson, Bell, and Pyrotect all sell excellent helmets. Generally with helmets you get what you pay for. Purchase a helmet that bears the latest "Snell" sticker and you will comply with all rules.

Follow these tips with your helmets:

1) When purchasing a helmet, make sure it fits snugly, but not too tight, as headaches may result. Keep in mind that your head will swell slightly on a hot day. The helmet should not be so loose with the chin strap off that it can rotate on your head while you try to twist it.

2) When adjusting your helmet before going out on the track, always tighten the strap as tight as you can stand it and press down on the top of the helmet to settle it on your head. Just like with seat belts, the chin strap will loosen up in a few minutes from the helmet settling.

3) Use goggles or a shield to protect your eyes from dust and flying debris, but use caution when using a full face helmet and a shield. In a enclosed car, the driver may not get enough ventilation to replenish the oxygen supply inside the helmet. This situation is worsened by the use of a neck brace or head sock.

4) Do not paint, sand or drill on a helmet. Some paints will attack the fiberglass, while sanding and drilling will weaken the structure. Let the helmet manufacturer do any painting or drilling.

Driving Suits

Nomex has long been the standby for suit material. This material is made by Du Pont and all the safety manufacturers buy it from them. While Nomex has excellent heat retardation capacity compared to cotton or even "fire proofing", and it has saved many drivers from severe, if not lethal burns. Nomex has the nasty habit of shrinking under heat, then tearing apart. After the material has torn it will leave the skin exposed.

In a 245 page report entitled "The Transient Thermo-mechanial Response of Protective Fabrics to Radiant Heat" published by the US Air Force in May 1977, the Air Force states: "...if the suit shrinks excessively, the protective layer of still air between it and the skin or between it and the inner garment is destroyed allowing the further transfer of heat directly..."

Nomex can be dyed, making the fabric more attractive to the fashion concious driver. Petroleum based dyes are used in coloring Nomex. These dyes are not fire retardant and, in fact, contribute to combustion of the fabric.

Safety equipment manufacturer Pyrotect has pioneered a relatively new fabric — Polybenzimidazole — (PBI for short) that offers superior thermal protective performance and virtually eliminates shrinkage and embrittle-

The Pyrotect PBI 3-layer driving suit is cool even in the hottest weather and it provides the driver excellent protection. A good helmet, gloves, shoes and underwear are good safety insurance. The backing on these gloves is of Nomex material and is dyed red so hand signals can be seen clearly. A visor or goggles should be worn even in a closed car to prevent eye damage.

ment when exposed to flame. PBI (a product of the Celanese Corporation) only comes in one color — gold — as the fabric will not take dyes. However, suits made of PBI are attractive and PBI offers another advantage over the other fabrics that closed car drivers will enjoy — PBI is much cooler to wear than a similar suit made of Nomex or Durette.

Underwear

Fire retardant underwear (socks, Long Johns, long sleeve tops and full face hoods) will add several seconds of safety if the driver is exposed to fire. Areas of the body such as the neck, ankles and wrists are often burnt when the driver does not wear fire retardant undergarments. Full underwear is highly recommended to drivers using a single layer suit. However, a multi-layer suit and underwear will greatly enhance the chance of escaping injury from fire and heat.

Gloves and Shoes

Gloves are important in a fire to allow the driver to remove his equipment befor bailing out without burning his hands. If the driver burns his hands, he will not be able to release his harness or even get out of the car. Gloves are also important in road racing to prevent blisters from the gear shift knob.

The same goes for the shoes — the driver will have a hard time escaping a burning car on blistered feet. Some cases have been reported where the drivers could not apply the brake because their feet were too badly burnt after a small fire broke out in the engine compartment.

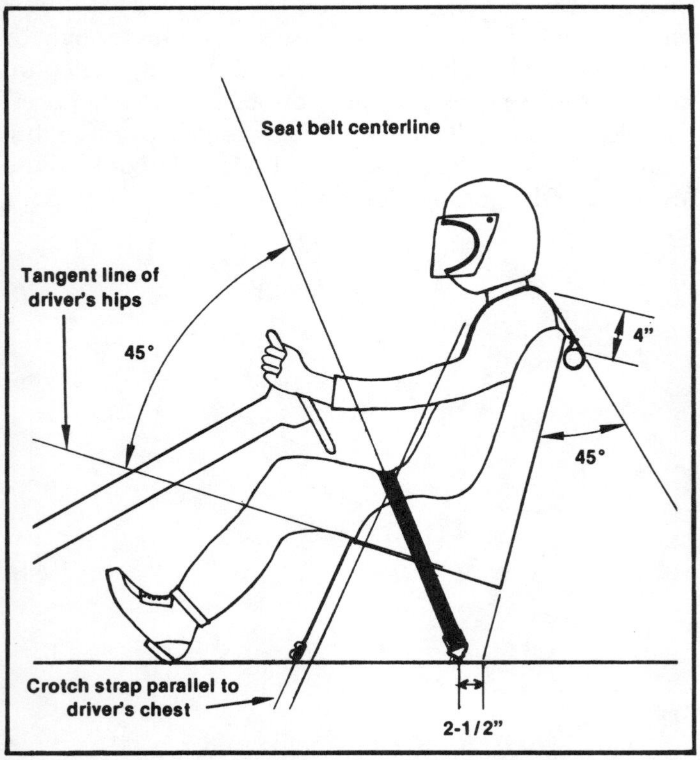

The correct mounting locations for the driver restraint system.

Belts and Harnesses

Belts and harnesses must absorb a tremendous amount of load in accidents. To be sure of getting the best for your money only purchase belts from reputable US manufacturers, such as Pyrotect, Simpson or Deist. Some foreign harnesses are plenty strong, however, the buckles cut the fabric well before the maximum allowable forces are encountered.

One of the most common mistakes with harnesses on stock cars and sedans is that the installers use much too long of a belt or harness. A 3 inch belt can stretch as much as two inches per foot in a good size crash. Head tilt, neck stretch, slack in the harnesses and 2 inches per foot of belt stretch can easily cause a driver to hit the steering wheel, or worse, the driver can slip out of the harnesses altogether.

Another thing to avoid is installing the shoulder harness mounts too far from each other. This will cause the harnesses to slip off the driver's shoulders during a race.

Watch the angle at which the lap belts pass over the driver's thighs. The lap belt should be at about 45 degrees from the driver's spine and the mount should be on a plane with the center of the driver's torso. The shoulder harnesses should pass over the shoulders at a 90 degree angle to the spine. The "Y" type strap harnesses are not SCCA or USAC approved but these are very popular with the dirt racers. The double harnesses should mount behind the driver from four to six inches apart with a short length of strap.

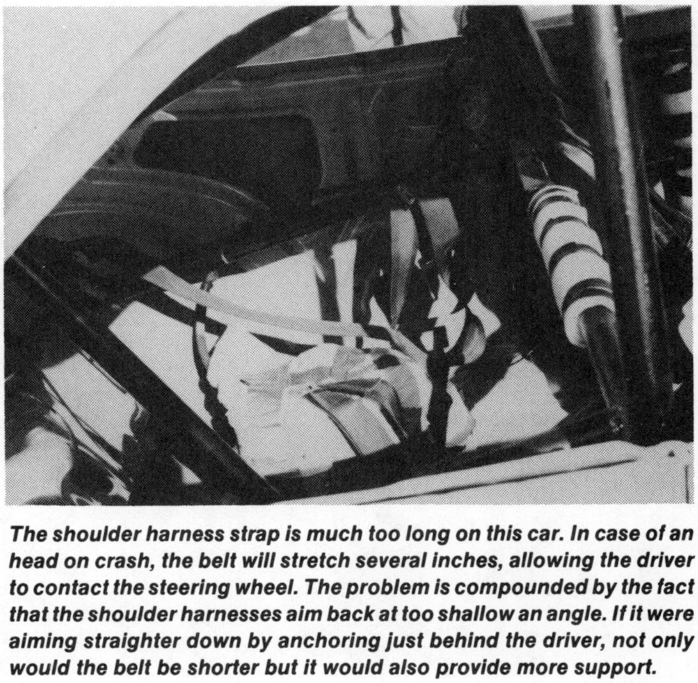

The shoulder harness strap is much too long on this car. In case of an head on crash, the belt will stretch several inches, allowing the driver to contact the steering wheel. The problem is compounded by the fact that the shoulder harnesses aim back at too shallow an angle. If it were aiming straighter down by anchoring just behind the driver, not only would the belt be shorter but it would also provide more support.

Neck Braces

A neck brace will not only absorb some of the movement generated by the head in an accident, but it will also relieve some of the strain on the neck stemming from the "G" loads in corners. The full face helmets are notorious for breaking collar bones. What happens is that when the car hits something, the car slows down but the driver keeps moving.

The body is held solidly in place by the seat and the restraint system, but the head, with the added weight of the helmet, will pivot about the neck and the lower lip of the helmet contacts the collar bone with enough force to break it. A good neck brace will cushion the shock to help prevent injury.

Fire Systems

A "fire system" ranges from a basic hand held fire extinguisher to extensive semi-automatic, remotely operated fire suppression and life support systems. The Phoenix fire suppression system we used on our project car consists of a five-pound bottle for the engine compartment and an eleven-pound bottle for the complete car. Each unit is independently operated by remote knobs at driver's easy reach.

In case of an engine fire, only the small unit can be set off, while in case of a major fire both bottles can be used. Most race car fires occur in the engine compartment. Using up a system that covers the whole car just to quench an engine fire would leave the driver without fire protection for the rest of the event. Follow manufacturer's instructions when installing a fire system.

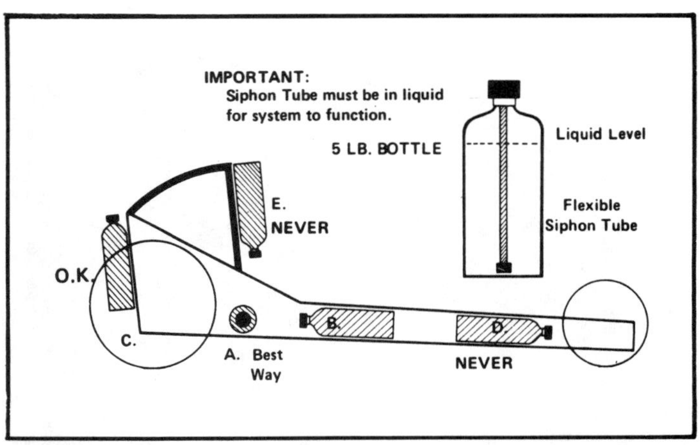

These fire extinguisher bottle mounting tips are courtesy of Specialty Equipment Manufacturer's Association.

Miscellaneous Items

Window nets were designed to restrain the driver into the car if he should slip out of his belts or should the harnesses stretch enough to allow his body or arms to protrude through the window. Another function of the window net is to keep parts and debris out of the car (loose wheels, fiberglass, etc.).

While the window net is intended to improve safety, it can itself be a hazard. Be sure the net is easily detached to allow quick exit in the event of a mishap. The heavy webbing nets can also be a visibility nuisance. These nets still provide good visibility when looked directly through, but at a sharp angle, such as from a rear view mirror, the webbing will block all visibility. Never use tie straps, wire or any other material to hold the net. These would not sustain a heavy blow or they would restrict quick exit.

Driveshaft loops are required by most associations and clubs. Loops should be manufactured out of ⅛-inch by 2-inch steel strap and located at both the front of the shaft and the rear. Some triangulation will help the loops from bending under the pounding from the shaft. Whenever weight permits, it is advisable to build a full length tunnel around the driveshaft.

Transmission/flywheel shields are strongly advised, even if your particular association does not require one. These blankets can be purchased commercially from companies such as Pyrotect or they can be fabricated out of ¼-inch steel plate.

A fuel pump shut-off switch, such as Ford Motorsport part #M-9341-A111, will turn off the fuel pump if the oil pressure dies. This unit should be given serious thought by road racers or long track oval trackers who might encounter oil pressure drop for long periods of time. Fuel pressure loss could stall an engine in a dangerous place, but blowing a motor from lack of oil pressure could cause even more dangerous problems.

Chapter 2

Rules

In recent years, rules in SCCA and IMSA have been modified to the point where a short track stock car can be run competitively in events such as SCCA's GT1 through GT5, Trans Am and IMSA's GTO and GTU categories. Building a car to run both Mini Stock short track and, say, GT3 SCCA club racing, involves some compromises.

The major areas where rules differ are: wheel and tire widths, engine setback, body modifications and racing weight. Other rules apply in SCCA and IMSA such as brake lights and use of the rear body panel versus the open rear end on short track cars. Road racing associations have regulations regulating aftermarket parts, transmissions, heads and carburetion rules. Fortunately IMSA and SCCA rules are used nationally, a lesson to be learned by Mini Stock associations.

The association under which our Mustang runs in Mini Stocks allows a 6-inch engine set back, while SCCA does not allow any setback, and the first spark plug must line up with the upper ball joint. To be competitive and legal in both GT3 and Mini Stock, our car would have to have two separate engine locations, meaning that the radiator would have to be relocated, motor and transmission mounts would require two positions,and the car would require two different length driveshafts. Running endurance races will involve installing windshield wipers, headlights and instrument lights.

While most Mini Stock associations rules state "stock appearing" bodies only, IMSA and SCCA have strict body shape rules. In 1986, for the SCCA stock roof, door posts and windshield must be retained and the door posts must be at the stock angle. For IMSA GTU, the body would have to be approved before the car is allowed to race unless an already approved body is used.

Mini Stocks can run front or rear engines. Rear engine cars must use a fully independent rear suspension system, which reduces unsprung weight and allows to set the camber in the rear wheels. This car was track champion using a Volkswagen motor and Ebeling chassis. Note the side rub bar, side skirts and the Streaker rims. Even though aluminum bodies look better than fiberglass ones, we chose not to use one on our car because of the difficulty in repairing wrinkled aluminum. The right rear fender on this car is evidence of our decision.

Speed costs money. Some racers are still very competitive in unibody cars with an integral roll cage built-in. The major drawbacks to building a car this way is weight and the chassis rigidity will not be as high as with a tube chassis.

Chapter 3

Chassis, Body, Graphics

The frame and chassis structures perform some functions that most people never consider. The whole chassis performs three basic functions, no matter what kind of car it is under: 1) the chassis is a mounting point for all the components necessary to operate a racing car, from the engine to the smaller components like the oil filter.

2) The chassis should be designed and built as the backbone of the car so it can transmit loads to and from the tire contact patches, without twisting or deforming. If the chassis flexes it acts as a spring which is not a parameter that was designed into the suspension. Any weakness in the chassis will cause handling problems that may be impossible to solve without solving the basic design flaw. A strong, stiff chassis will make the suspension easy and accurate to tune.

The chassis should also be designed to handle the torque of the driveline without twisting or deforming and to maintain the correct driveline angle. The chassis should also provide some deformability in some pre-determined locations of the structure to protect the rest of the chassis in case of an accident. This is an important factor to keep in mind for the novice driver whose odds of crashing are higher than the more experienced racer. Some rigidity may have to be compromised in order to provide some deformability to the chassis. If a car constantly oversteers and spring and anti-roll bar changes do not improve the condition, chances are that the chassis is flexing too much and nothing short of a complete redesign will solve the problem.

3) The chassis must provide safety to the driver by supplying a strong, triangulated roll cage, side protection, and anchor points for the harnesses and seat hoop.

A nice advantage we have when building a chassis for SCCA or Mini Stock racing is that the legal weight limit is some 600 to 700 pounds over the race ready weight of the car. We had to add 650 pounds to our car to make the 2350 pound limit in our association. This fact gives us the flexibility of building the car without being concerned with weight. This is probably the greatest cost saving feature with cars of this type.

Engineering and Design

The complete frame and chassis structure must be considered when designing and building a cage. While the square tubing frame provides a base on which the cage is built, the cage itself supplies the triangulation to form a strong and stiff structure. A strong chassis will give predictable weight transfer in the turns and it will make the suspension tuning more accurate.

To get an idea how much a car flexes, place a long piece of masking tape level with the ground across the windshield and the back window, both at about the same level off the ground. Jack one of the corners of the car by a brake disc until another corner starts to lift off the ground. Look at the tape on the windows through the front or the rear. If the chassis flexes badly, the two pieces of tape will be crossed in an x shape. Try this with a passenger car first

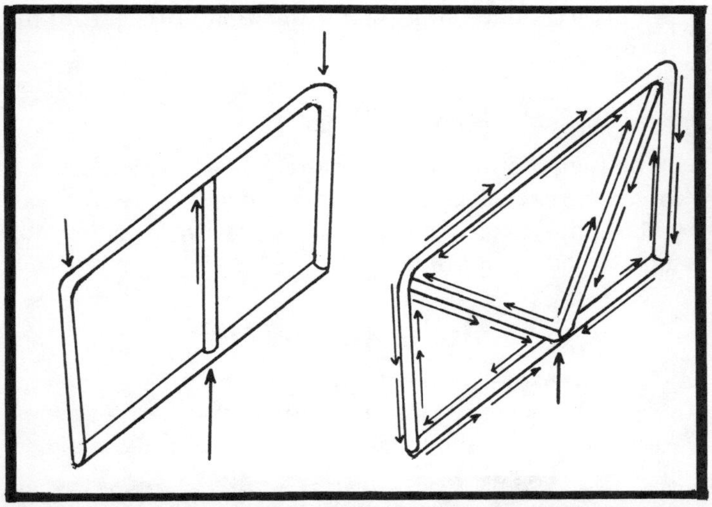

A comparison of two types of chassis structure, the left one untriangulated and the right one triangulated. The basic principle of triangulation is to spread the loads out evenly throughout the entire structure. In the right drawing, it is shown how a load fed in at the bottom center of the structure is evenly distributed throughout the entire structure. When the arrows of load distribution are pointing against each other on the same tube, this means the tube is accepting the load placed on it "in compression." In other words, the tube is literally trying to be compressed. This is its strongest asset. It resists compression and tension best, and bending the least. In the drawing at the left, all of the loads place the tubes in bending.

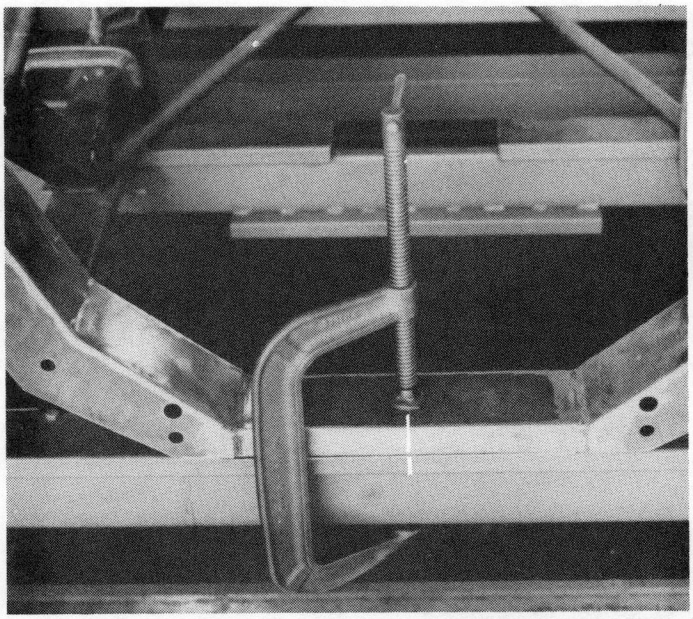

In order to get some accurate measuring points the chassis main rails must be built parallel and the crossmembers must be perpendicular to the frame rails. Here the c-clamp holds the front crossmember firmly in place while it is welded. Note the centerline scribed in by Ebeling that lines up with the line on the jig. These two lines are the cornerstone of the chassis.

Water levels are used extensively while constructing a chassis. Using a level ensures that all the bars are parallel to each other. The advantage to this is that if the car is hit, the chassis can be repaired easily by resetting all the tubes parallel to each other.

to get an idea of what bad chassis flex is.

It is preferable to design a chassis on sound engineering principles, rather than adding material later to the weak points. Triangulation is the most important principle to be used while fabricating a chassis. Everyone has seen the demonstration of the brick being held up by a simple piece of paper. By being folded, the paper is using its strength in compression. Another graphic example of good utilization of engineering principles is the concept used to hold shelves on their brackets. Without triangulation, the shelf would collapse, but as soon as a brace is added to the shelf, the whole unit becomes many times stronger.

A chassis should not have any tubes ending on another tube without a third tube supporting the first two tubes. When welding a tube to another, do so where the second tube joins another tube. Never weld a tube into a straight section without supporting the back of the section or that tube will bend under the load. Also the three tubes must meet at steep angles to take up as much load as possible without bending.

A diagonal bar should be used wherever a rectangular shape is formed by the tubes in the chassis. The diagonal bar triangulates the box, increasing the strength of the structure many-fold. This box shape is often seen inside the roll bar hoops and the engine bay.

Design Factors

There are five major parameters that affect chassis loading: 1) total weight of the car, 2) tire size and compound, 3) center of gravity height, 4) G loads developed by the car during cornering, and 5) power of the engine.

Total Weight of the Car

The loads are taken up by the coil-over mounts, the upper and lower control arm mounting points and the rear end locating device. If a car weighs 1,000 pounds and it develops 1 G of lateral force and its weight distribution is 50/50, the locating device in the rear will have to hold 500 pounds, and the same 500 pounds will transfer into the chassis and frame structure. Furthermore, the greater the G forces, the greater the amount of weight transferred and the greater the load on the shock mounts.

The loading points must be reinforced by using triangulation because they are the most heavily loaded points in the car. The heavier the car is, the more weight the chassis will have to support and the more weight will be transferred through its tubes. Heavier cars need more crash protection, because in a crash a heavier car has more weight behind it to crush the part of the chassis that has made contact.

Heavier cars are generally more powerful and the chassis must contend with greater torsional forces from the engine torque.

The left rear end lower strut mount is being readied to be welded to the rear axle tubes. Since the welding will warp the tubes, the rear end will have to be returned to the manufacturer to be straightened. Notice the level again.

C-clamps used to hold the parts to be welded must be very heavy to keep the parts from moving under the expansion created by the heat.

Tires

The bigger and stickier the tires are, the greater the G forces the car will generate. A car raced in a tire rule class will not need as stiff a chassis as a car with wide tires and sticky tread compound.

Center of Gravity Height

The lower the center of gravity is, the greater the cornering ability of the car is. The greater the G forces generated, the greater the loads that will be on the chassis.

End plugs like these are used to cap off the end of a piece of tubing. These plates also add strength to the welds and the junction of the two parts. Arrow points to the angle iron that runs the length of the jig to hold the main frame rail in place during construction.

Diamond shaped plates are added to welds to increase their strength. The plates introduce some metal in the tension plane.

G Loads

The larger the G loads generated by a car the greater the chassis stress will be. If we study the weight transfer formula $WT = \frac{CGH \times W \times G}{TW}$, where W is total weight of the car, G is the gravitational forces developed by the car, CGH is the center of gravity height and T is the track width of the car), we find that the G loads developed by the car are a multiplier of the weight transferred. Here we see that the track width also has an effect on the amount of weight that the car transfers, thus the amount of load placed on the chassis.

Engine Torque

The torque of the engine places a twisting force on the chassis. A weak chassis can flex enough under high power to change the angle of the driveline to a point of creating vibrations and u-joint failure.

The rear end also counteracts the torque that is applied at the rear wheels back through the chassis. That reaction can bend the chassis from front to rear.

This type of problem is uncommon with four cylinder cars, but it is every day occurrence with high powered cars like a Pro Stock dragster.

Chassis Material

The Ebeling frame is constructed of 2 x 2 x .090-inch 1040 mild steel at the front and rear stub, while the center section, under the roll cage uses 2 x 3 x .090-inch mild steel. Mild steel is much easier to work with than chrome moly which is harder to weld and to bend. Chrome moly is also more expensive than mild steel. The Ebeling main cage and chassis structure is formed out of 1 3/4 x 0.090-inch 1040 seamless mild steel. The smaller supporting tubes can be made out of 1 1/2 x 0.065- inch 1040 seamless mild steel, while smaller triangulating members can be made of 1 x 0.065-inch mild steel. These sizes are fairly standard in the industry.

Square tubing should never be used as an integral part of the roll cage as it does not have the strength of round tubing. Square tubing can be used in the frame and for things such as body mounts, bumpers and jacking points.

The rear clip main rails are clamped down to the jig and to the sides to the angle. This locates the pieces in both planes and makes for a square rear clip.

The extra length of tubing at the arrow is retained to protect the battery. Note the round tubing in the background forming a double V pattern. This shape strengthens the floor and prevents the chassis from flexing laterally.

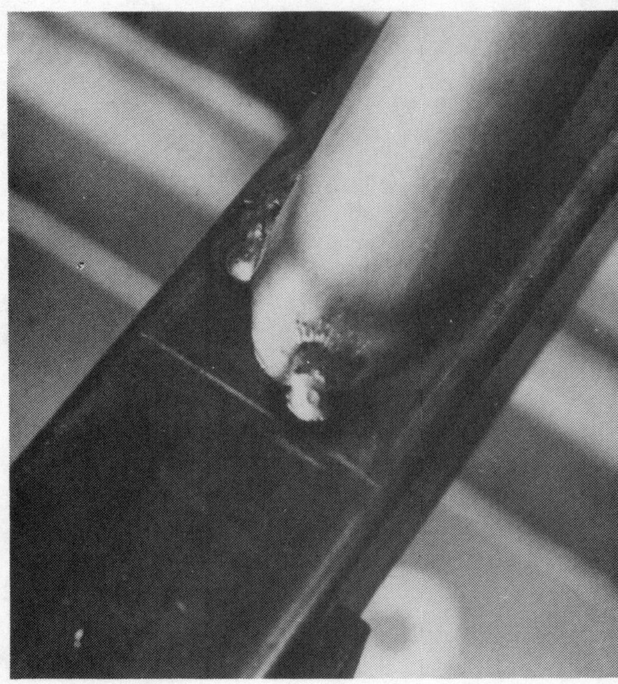

Before welding, the tubes are fitted as tight as possible to the part they will be welded to. The parts are first tacked in place and then are welded in.

Stock car chassis are MIG welded. Wire feed is fast, strong and it can fill gaps in the joints that other forms of welding would take a long time to weld.

On the jig our chassis looked like a tank. Once on the ground it looked low and light. Mini Stocks with the Ford 2.3 weigh around 1600 pounds, dry. The chassis could be reinforced, but Ebeling uses some sound engineering principles to strengthen the chassis.

This bar triangulates the fuel cell hoop. If the car gets hit in the rear, the hoop will not fold into the tank.

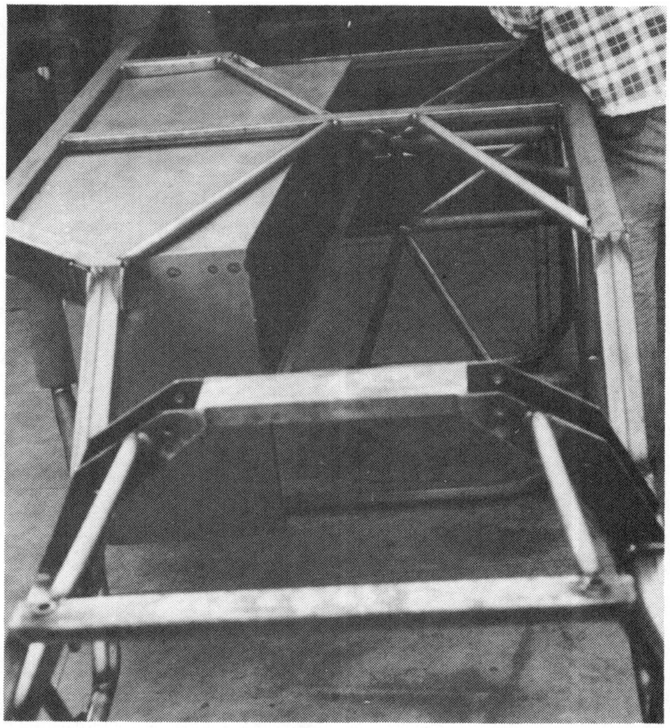

The chassis are welded in two steps. First the top welds are done on the jig, then the chassis is turned upside down and the bottom welds are finished. This way overhead welding is avoided.

The third bar mounts on this bracket. The bracket is mounted directly to the chassis. Some other chassis builders use additional bars to run the mount back, adding weight and cost to the chassis.

Tube Bending and Welding

The tubes should be bent on a mandrel to avoid deformation. When a tube is bent, the outside of the bend actually stretches the material and if the inside of the bend is not supported properly it may collapse or wrinkle. This causes the bend to lose its strength, causing a general weakening of the structure. SCCA and IMSA require the tubes in the roll cage to be bent over a mandrel and not to have any wrinkles and signs of collapsing.

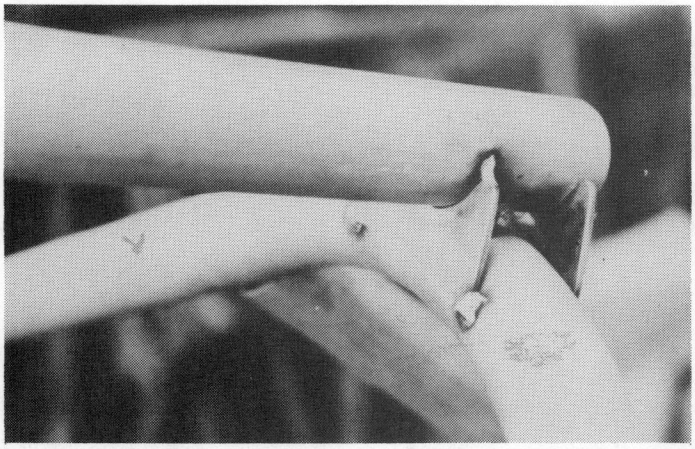

This method of mounting the front hoop of the roll cage adds strength to the weld and the extra height yields a bit more head room in the driver's compartment.

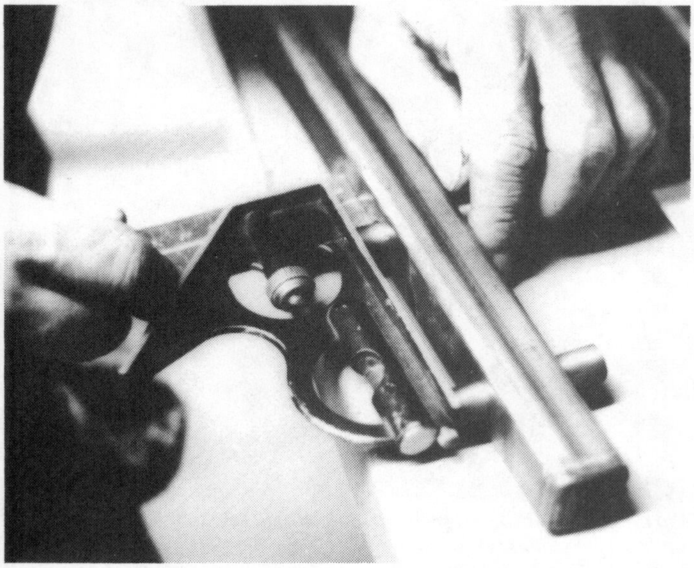

Whenever running tubes through a structural member, always weld both sides of the hole. Here the tubes are being aligned and set to height using a square.

It is not true that bends weaken a chassis, as long as the bends are not in a stressed section of the chassis, or as long as it is triangulated.

The welding should be done by a qualified welder. Wire feed (MIG), heliarc (TIG) or arc welding should be used on the chassis and frame. Ebeling chassis are wire feed welded because of the ease and rapidity with which the joining process can be done. However, Ebeling warns that a wire feed weld can look good, but it may not have the penetration needed to make a strong weld if the welder is not well experienced. Heliarc can also be used on a chassis and cage. The only drawbacks to TIG (Tungsten Inert Gas) are that the welding takes longer than MIG and TIG requires tight, clean fits between the parts to be welded. TIG requires a more experienced welder to do. Arc welding is also OK if the operator is competent and if he uses the correct rod and amperage setting.

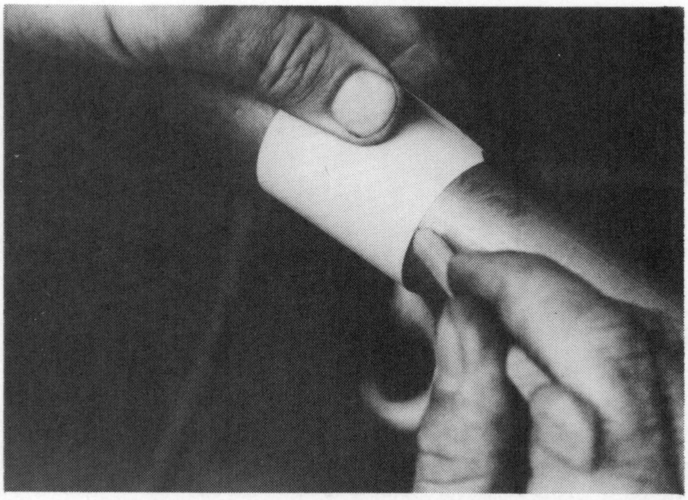

Using a piece of thin cardboard wrapped around a tube allows a line to be drawn around the tube before drilling the holes. After the centerline is drawn, the holes can be drilled to insert the small mounting tubes.

Installing the Body

The body doesn't have to be one of the last things to go on the car. In fact, we found the body installation to be quite messy to the rest of the car because of the fiberglass work. The car needs to be at ride height and with the same size wheels and tires that are going to be used for racing. Speaking of tires, watch out for the rivet studs—they make nice holes in new tires!

Basically the trial and error method seemed the most simple method of installing the body. Use blocks of wood, dowels, wood stakes, c-clamps or anything you find convenient to hold the body parts up. A couple of friends will also be helpful to "eye-ball" while the others hold the body parts in place.

We used the Mini Mustang body from Stock Car Fiber-

The very first step in installing the body is to lay the hood on the front end and Cleco the fenders on. The fenders should cover as much of the tires as possible. Use small c-clamps to get the hood and fenders in the ball park before drilling for the Clecos.

Next install the roof. Find the center of the hood and line it up with the center of the roof. Use a plumb bob to center the rear of the roof in relationship with the frame.

We used this method to hold up the rear of the roof during construction. The roof should be parallel to the frame.

The front part of the hood can be added next. Be sure the front wheels are centered within the wheel wells.

glass, and it fit easily on the Ebeling chassis. Gary Ebeling has also installed some Stock Car Fiberglass Mini T-Bird bodies and they are even easier yet. The advantage to the T-Bird body is that the doors and rear fenders are one piece.

I do not see how the body installation can be done without Cleco fasteners. Clecoes are fake pop rivets that are installed in the rivet holes with special pliers to hold

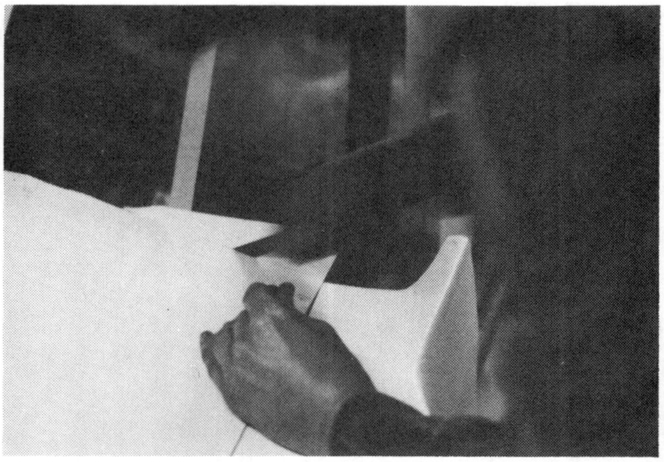

If the hood needs to be moved back on the car, trim some of the fiberglass in front of the windshield

We made these little props on the left to help install the body. The stud on the right simulates the hood pins to help set the height of the mount.

the fiberglass temporarily in place to allow fitting of the body panels. Clecoes and pliers can be purchased from Vic Irvan Enterprises.

Body Fitting

The first thing to do is install the sheetmetal that the lower part of the windshield fastens to. If your chassis did not come equipped with these parts, the easiest thing to do is to take some measurements off another car or send away for the pieces from the chassis builder. Ebeling can fabricate these pieces for the Stock Car Fiberglass bodies.

The next step is to set the roof in place. To keep the stock look, set the roof on blocks. Take some measurements off another car using the same body you are using. Measure the height of the roof in relation to the roll cage to attain the desired roof angle. Cleco the windshield posts in place.

Next, set the hood, front fenders and grille over the front end on some blocks and Cleco the fenders in place, making sure the hood covers the fenders evenly on both sides. Allow enough room for the air filters if you run downdraft carburetors. We used a floor jack at each side of the front of the hood to get that height set, while under the hood we set some adjustable stands we fabricated out of 1/2 inch steel strap and wood dowels. The strap was bent at one end and attached to the chassis tubing by a hose clamp. At the other end we attached the wood dowels with hose clamps, so they could be moved up and down to set the height we needed to support the panels. Allow plenty of room for the wheels to turn and bump. The fender wells will have to be trimmed later. Cleco the hood, fenders and front end of the hood together. Get the wheels centered within the wheel well.

Using the same method of supporting the panels and eyeballing, you can now Cleco the hatch and rear quarter panels. Getting the whole thing straight is done with the doors. Set the doors so the crease in the sides forms a straight line with the crease in the hood and the rear fenders.

We did not watch the roof angle close enough. This was the result.

The quarter panels are lined up with the contour on the doors. If the doors are parallel to the frame and the roof is parallel to the doors, the quarter panel will match up with the roof. Notice that the top of the hatch slips under the roof and over the quarter panels.

Details of the hatch to the quarter panel fit. The rear bumper was installed to help fit the rear part of the body. This piece is required when racing in SCCA. Remove the rear bumper fiberglass when possible to help get rid of the under-hood air.

The arrow at the bottom points to the lower body support. Top arrow points to a temporary body support used while the body is being fitted. This body is a T- Bird.

Body Supports

Support the body where needed. The hood should be supported on each side in front of the windshield, at each side just above the headlights and on each side near the lower part of the door. To support the hood use Mr. Gasket hood pins, part number 1217. The roof gets riveted on the lower part of the windshield pillars and on four supports on the roll cage. The doors are riveted to the rear fenders, the windshield pillars and to the sheetmetal on the bottom of the left door and the square tube that was run along the lower part of the right side of the car to

The wide strap at the bumper will keep the bumper from hooking up with other cars on the track, and it provides an ideal place to rivet the rear fenders.

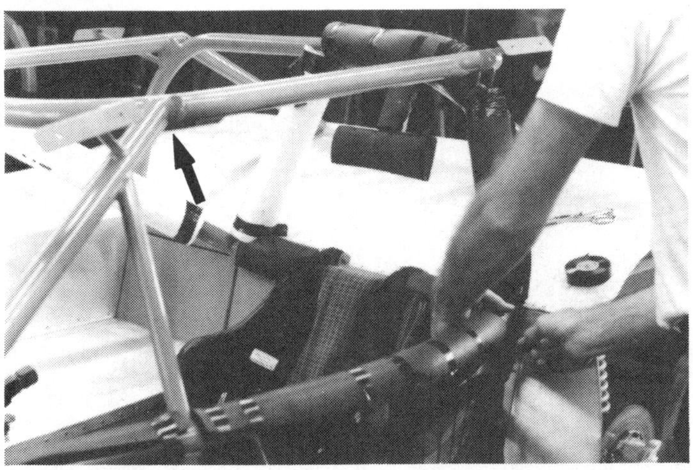

There always seems to be an after thought about just finished race cars. Many people race the car a few times before they paint the chassis. This car is being reassembled after painting. The second assembly is a good time to check everything. The padding is easier to install with the body off. The small vertical bar at front of the cage would be more effective if it was welded to the junction of the roll cage hoop and the forward brace. Note roof mounts on the roll cage.

support the body. The hatch is riveted to the top of the roof, the quarter panels, the top of the fenders and the top of the sheet metal under the back window. The rear fenders are riveted to a 2-inch strap that extends from the ends of the rear bumper and to a support that comes up from the bars running from the top of the roll bar to the rear of the frame. The rear fenders are also supported by the sheetmetal that separates the driver's compartment from the wheel wells.

To rivet the body, only use aluminum rivets. Steel rivets will rust. Aluminum body rivets and rivet guns are available from Vic Irvan Enterprises.

Graphics and Color Design

Having a professional looking car is important for several reasons: 1) You will get recognition for yourself and your sponsors as a professional and you will get more press and track announcer coverage than the other drivers will. 2) You will have a better chance of attracting a sponsor than you would with an unprofessional looking car. 3) In addition a professionally finished car will photograph better at speed, increasing your chances of getting newspaper coverage and track program coverage. The car will also be easier to keep track of by the spectators during a race. 4) Pride. You probably spent many months of hard and expensive work on the car, so you might as well show it off at its best. Feeling good about your car will also improve your self confidence on the track.

There are a few rules that need to be followed to get a professional looking car. An excellent source of information on this subject is Gary Smith's book, Race Car Graphics, The Professional Touch, also published by Steve Smith Autosports.

Cal Smith doing the masking on the car. The graphics were first painted on a model car to give us an idea of the way the finished product would turn out. The car is blue with a red stripe and the upper part of the body was left the original white. We found the unpainted portions to be difficult to keep clean.

We kept the graphics flowing with the body lines of the car. The red stripe was kept in proportion with the size of the car. A bigger car could use a wider stripe.

Mark the outline of the windows by laying a pattern on the actual piece of plexiglass under the opening it is to cover. With a pair of dividers, mark the extra width you will need to install the piece in its place.

We washed the whole body down with rubbing alcohol and then sanded it with 320 grit sand paper. The body was washed down again with rubbing alcohol and then painted with Imron paint. No primer was used and the results were excellent. Remember to wear a respirator when painting because overspray is highly toxic.

Basic Rules

With race cars there are two kinds of graphics that must be dealt with: 1) characters, which refer to the letters and numbers placed on the car. This includes their shape, size, color and positioning on the car as they relate to the rest of the car. Legibility and ease of capturing the characters on film must also be considered. 2) Shape graphics refer to any stripes, or patterns of colors applied to the car.

The shape of the numbers should be kept simple to make scoring and race reporting easier. A simple number design will blend in with a greater variety of paint schemes you might want to choose for your car. For instance, if you use an antique Gothic style on your numbers, you should match the rest of the lettering on the car with the same lettering style. A whole race car lettered in Gothic would be hardly readable standing still, much less at speed.

The stripes, patterns and colors applied to the car must be chosen very carefully. Stripes should follow the peak lines, the cut lines, and the shoulders of the body and its design lines. This method is evident on our car and it is explained in detail in the aforementioned book.

The colors should be chosen so they complement each other. A complementary color chart can be obtained at paint stores. Complementary colors are the opposites: yellow and purple, red and blue etc. To see how your car will look, build a plastic scale model of it and paint it the way you would like to paint the race car. A trick used by ad agencies is to build a scale model and place it at a distance relative to the scale size of the model to observe the

The logos and lettering was done by Auto Signs Unlimited. The vinyl decals were easy to apply and looked as good as paint.

We studied the logo positioning carefully. However the Ford logo should have been in the place of the Steve Smith Autosports loge so as not to be so distorted by the hood scoop. The roof kept its original white color so we could promote this book on it. Notice that the typestyle is very simple to make it easy to read even at speed. So many racers make their lettering too small to be read from the stands.

The stripe follows the shape of the body. We tried to keep the wheel well from cutting into the white stripe so as not to cut up the lower part of the paint scheme. The decals are all parallel to each other and to the stripes on the car. The shape of the decals dictated their positioning. If you get a sponsor and want his logo enlarged, send it to Auto Signs Unlimited for reproduction (like the Ford logo on our hood).

The number is large, bold and, most important, easy to read. The stripe accentuates the paint scheme as it turns up above the rear wheels.

The rear of the roof needs to come up some, but the finished car still looks attractive, just the kind of car a sponsor would like to hang his shingle on.

Look at the difference in contrast between the numbers on the door and the unfinished numbers on the roof. This car could use some graphics to make it stand out more.

legibility of the designs and lettering. For instance, if the model you built is a 1/10th scale and the spectators at your track stand about 30 feet from the track, place the model three feet away and look at it as if you were looking at the real car.

When we designed the basic color and graphic scheme on our Mustang, we went through magazines and found the cars we liked the best. Three things stood out: 1) all our favorite cars had colors that complemented each other, 2) all the cars had lines that flowed with the body, and 3) the cars had graphics that were easy to read and that were highlighted or outlined to cause them to stand out and blend in with the rest of the car. Our Mustang looks a lot like Terry Labonte's Piedmont Winston Cup car. That's OK because we will never race this car on the same track with Terry. But use caution in getting ideas from cars at your track and especially in your class so you won't look like a copy cat. One major difference between our car's scheme and Labonte's car is that our roof was left white so we could use the space to promote this book on it.

When designing the scheme on your car, take the overall size and shape of the car in consideration. The paint

The nose piece of this Mustang was replaced with vinyl after it had been hit a few times.

scheme on a small stock car will be different than the designs on a formula car or even a Grand National car. Also, if you have a major sponsor, you will have to blend in his colors and logo with the whole scheme of the car.

Stock car racing is very functional. This aluminum siding, for instance, holds the front nose piece in place and it can be replaced quickly and cheaply. The front nose piece is flexible to take up the rigors of short track racing.

This brace can be used to support the sides of the front nose piece if it is made of vinyl.

Fiberglass is easy to repair and for a short track car that is an advantage. Aluminum is much more difficult to pound out and an inexperienced person will end up with a bunch of wrinkles.

A rotary sander is first used to rough up the glass and then the mat and resin are applied.

Chapter 4

Mounts And Brackets

When we started the Mustang project, we assumed that the chassis would come with all the mounts and brackets already on. Gary Ebeling, our chassis supplier, explained that most mounts and brackets are customer preference items. Some brackets such as the radiator, oil cooler and fuel pump mounts are based on the make of the item to be mounted.

Other factors influence bracket placement such as driver's size, equipment used, form of racing (road course or circle track), rules, engine placement etc. The chassis builder could build you a complete chassis with all the brackets and mounts, but he would need to know (or preferably have in his possession) the parts to be mounted, and know such things as engine location (offset, set back, front or rear location).

One of the functions of a chassis is to provide the builder with a place to mount accessories such as the fuel pump, filters, the seat, window net, etc. Other items such as the motor and transmission mounts need a place to be welded to.

Chances are that you are not so fortunate to know how to weld and to own your own TIG or MIG welder. Then your best bet is to either enroll in a welding course at your local college, enlist the help of a friend who has the needed skills, equipment and/or, hire a welder to come by your place a couple of times a week and do the welding for you. Whether you do the welding yourself or you hire someone to do it for you, here are a few tips to make the job better and easier.

Metal Gauge

The motor mount plates, if you use that type of mount, could be 1/4-inch, while most low stress mounts and supports can be made of 3/16-inch plate, strap or angle.

Choosing The Metal

If you have to pay a welder to do the work, you will

The mounts can first be made out of cardboard and then transferred to steel.

definitely want to prepare the metal so it is ready to weld. Avoid rusted or painted metal. Also some steels come with a lot of mill scale that makes welding and prepping very difficult. Buy a clean cold rolled steel. 3140 is expensive, but it is easy to weld and to prepare. 3140 rusts very slowly so it can be stored for a long time. A good metal dealer will be able to advise you on your special application. Your welder might have a preference, so check with him.

Aluminum

You will not make any brackets out of aluminum but some welding to the radiator, catch cans or oil sump tank may be necessary. Not all aluminum is weldable. Tempered aluminum (denoted by the "T" designation) does not weld well. The most common and readily available aluminums that are also weldable are 3003, 5052 and 6061. 3003 is also easily formed and it can be used for interior sheet metal, tabs, mounts and is easily machined.

Purchase aluminum in the "O" condition if it is to be welded. When buying threaded bungs or fittings to be welded, get unanodized aluminum. Anodizing does not weld and it would have to be sanded off completely.

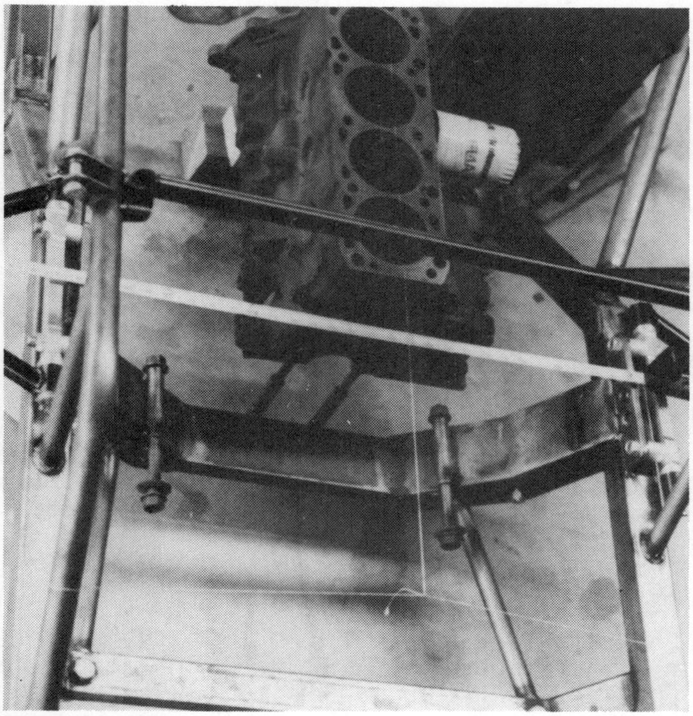

Before starting work on the engine and transmission mounts, install the oil pan and, by using a straight edge across the bottom of the pan, measure the distance from the bottom of the pan to the center of the crankshaft. Next set the chassis at ride height in front and level it front to rear, and side to side. Install the engine in the car, minus the pan, and support it with blocks. Level the motor and set its height with the pan clearance about 1/2 to 1 inch above the bottom of the frame. Here a method of centering the engine is shown. A string is run straight back exactly at the center of the frame and the center of the engine is aligned with the string. Before going too far with the mounts, install the transmission to be sure its tailshaft will clear the center crossmember.

Aluminum cannot be MIG (wire feed) welded. It must be heliarced (TIG). The aluminum to be welded must be perfectly clean, preferably sanded with fine sandpaper. Do not clean the metal with a wire brush. Impurities will enter the metal and form little craters when it is welded. Use a new stainless steel brush if you have to and wipe the part down with acetone and let it dry.

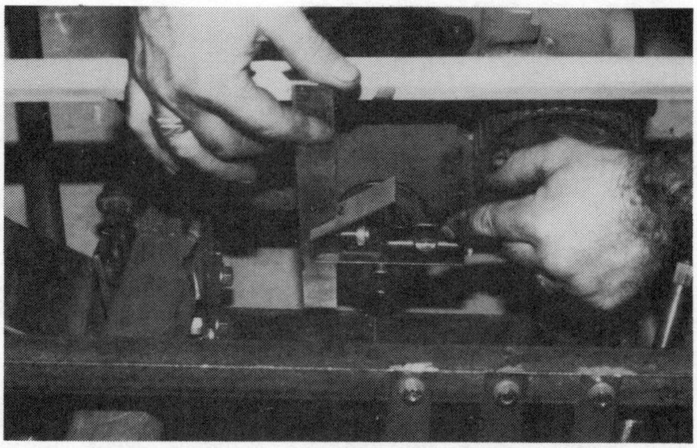

If the chassis builder left a centerline on the front cross member, use it as a reference to center the engine. Here the center of the crankshaft is found by aligning the rule with the center of the crankshaft and the centerline on the frame. The water level will keep the rule vertical.

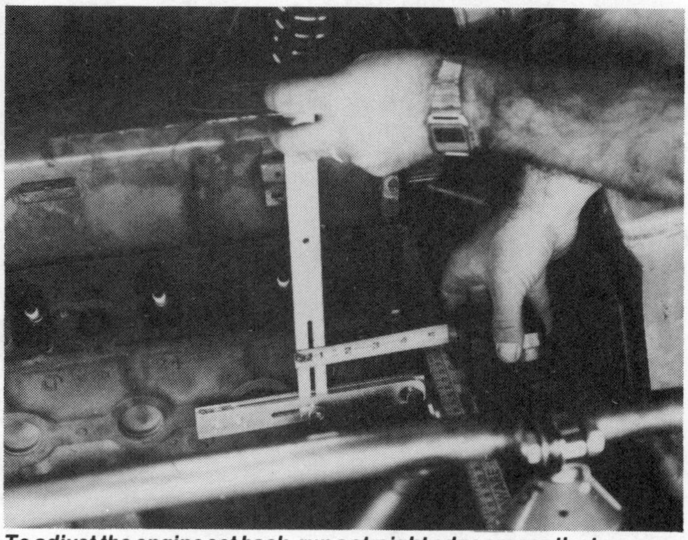

To adjust the engine set back, run a straight edge across the two upper ball joints. Then measure back to the number one spark plug. We found it easier to use a rule set against the side of the head and measuring off it. You can now start taking measurements for the mounts.

Metal Shape

1.25 inch by 1.25 inch angle can be used for the fuel tank mount, radiator, oil cooler, steering support, fire bottle knobs and dry sump tank mount. Tubing 2-inch by 1-inch by .060-inch wall can be used for the fuel cell mount, ballast mounts etc. Steel strap 2-inch by 1/8-inch can be used for body supports and miscellaneous tabs and brackets, while strap of 1-inch by 1/8-inch can go on body supports and radiator mounts.

When measuring for the motor mounts allow for clearance for the oil pan, the steering rack and its shaft, the steering mounts and the crankshaft pulley. Also keep in mind that the mounting bolts will need easy access. Here the left side mounting bolts are too close to the steering rack mount.

These centering pins are easy to make out of old bolts. Just saw off a bolt the size and thread pitch you need and insert it in the bolt hole. Next, place the part you want to mark (in this case the motor mount) and tap it with a hammer to leave a mark where the hole can be drilled. For better accuracy, first drill a couple of holes and reinstall the part and tap the rest of the guide holes in. Enlarge each hole one or two drill sizes higher and you will have a part that is easy to install and that will be reliable.

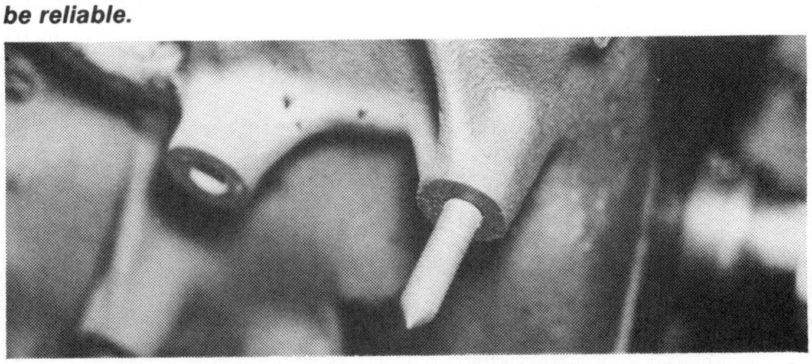

The two top mounting bolts of the right side motor mount required a shim between it and the block. The centering pin we made for this hole had to be longer than the others. Use a grade eight bolt if possible to make these dowels. Grade eight bolts are much harder than a lower grade bolt and will make a better mark. Also be sure the point of the dowel is in the center of the bolt or the hole will be offset.

This shows the details of our right side front motor mount. Notice the dry sump pump mount and the location of the lower mounting bolts. They are located in an easily accessible area and away from the crossmember. The two spacers at the top can be made of heavy wall tubing. It would be a good idea to safety wire the fasteners that run into the block. If the radiator hose is going to come close to the pump mounting bolts, use round head Allens to avoid cutting through the hose.

Details of the front mounting plates on the crossmember. Plan ahead by drilling the holes in the mount and the plate at least 2 inches above the crossmember to allow for fastener installation. Before welding the plates on, c-clamp them on the mounts, locate them on the crossmember and then remove the mount and the plate for drilling.

You may want to use 2-inch by 1/4-strap for the motor mounting plates on the frame. Half-inch square tubing will go on the lower part of the body, hood pin supports and, this stuff makes nice radiator and body mounts. One-inch square tubing can be used for bumper guards, jacking points and pedal mounts. Use .060-inch wall thickness.

Which Form Of Welding

Stock car chassis are almost universally wire feed welded (MIG). Wire feed is fast, strong and it can fill large gaps. The MIG bead is not always pretty, but it gets the job done. TIG (heliarc) can be used on small brackets and mounts that do not require long welds. Heliarc is nice to use on bench-welded parts and it is necessary on aluminum. TIG does not fill gaps too well and the metal to be joined must be perfectly clean, inside and out. Brazing works well on dirty metal and around joints such as water outlet elbows, temperature sending units and when running a small tube through the chassis main structure.

Metal Prepping Tips

When fabricating a bracket or mount, always take the welder in consideration and try mounting the part before welding. Remember that the welder may not be able to get his head with the hood on in certain areas or the welding tip may not reach far enough to weld your part on. Tack weld as much as you can first, drill the holes (if you could not drill them on before) and test your bracket. When a piece is MIG welded, allow plenty of room for the thickness of the weld if a bolt is to be used near the weld. TIG (heliarc) makes a smaller weld so it is not as critical as MIG when it comes to bead size.

Clean the outside of both parts to be joined before the welder shows up. Fit the part and mark exactly where you want it welded or tack it in place if you can do so. If you have to use muriatic acid before TIG welding, do so just before welding or the part will corrode.

General Fabrication Tips

The main idea is to think at least one step ahead of the game. Fabricators working with prints do not need to concern themselves with what will be mounted behind or in front of the part they are welding at the moment. But the lonely car builder must consider such things as: will this bracket or mount interfere with another component on the car, is the part mounted as low as possible and not be in danger of contacting the ground when the suspension bumps, will there be enough room to get a wrench and a ratchet on the bolts?

Also, the general layout of the car must be kept in mind. Why, for instance, mount the oil filter where it can be removed easily while that position would require angled fittings to be used. You may have to compromise on accessibility to maximize on function. On our car, for example,

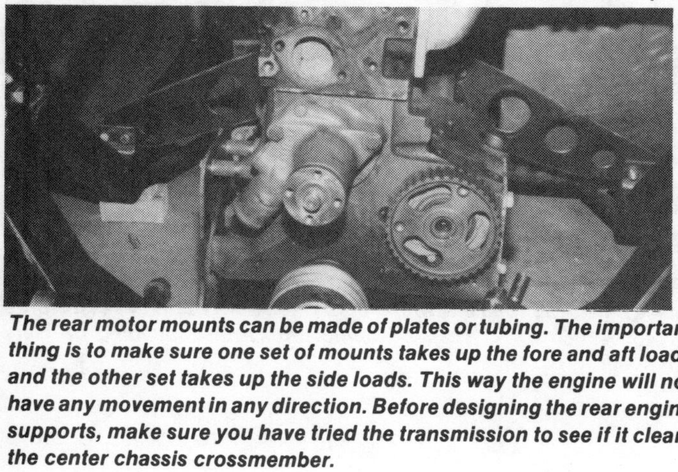

The rear motor mounts can be made of plates or tubing. The important thing is to make sure one set of mounts takes up the fore and aft loads and the other set takes up the side loads. This way the engine will not have any movement in any direction. Before designing the rear engine supports, make sure you have tried the transmission to see if it clears the center chassis crossmember.

we designed the oiling system components to be in line with each other.

Consider this check list when fabricating mounts and welding: 1) Safety, 2) Cost, 3) Strength, 4) Performance, 5) Crash worthiness, 6) Serviceability, 7) Weight, 8) Appearance.

This type of motor mount is easier to build than the ones made of steel plate.

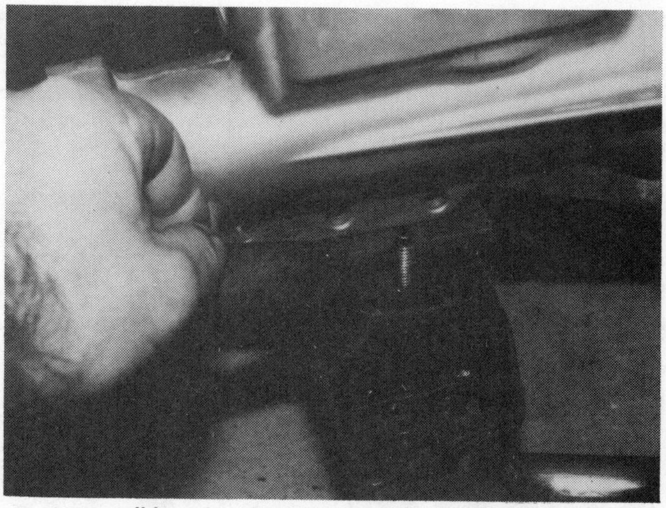

If you use solid motor mounts, you should rubber mount the transmission. The flexing in the chassis must be taken up somewhere. If it is not, it will break the bell housing or the transmission tail shaft. The transmission mount will have to be chosen based on the clearance you end up with under the tail shaft. We used a late model T-Bird mount at the tail housing.

Here is a list of most of the mounts and brackets which will need to be fabricated and or welded: bumpers, bumper mounts, radiator and oil cooler mounts, body mounts and supports, motor mounts, trans mounts, oil catch can, water overflow bottle, steering supports, dash, fire bottle knobs (do not mount the knobs on aluminum sheet metal) and bottle mounts, pedals, ballast, fuel cell, rear end breather, steering wheel quick release hub, oil and fuel filters, cross bar in front of engine on top of the chassis, battery box and ground, master kill switch, interior sheet metal supports, jacking points, jacking bar, rear view mirror, window net, driveshaft hoop, water outlet elbows, fittings for pick-up of coolant temperature gauges, gear shift lever, dry sump pump mount, brake caliper mounts, sway bar mounts, side contact bars, seat mount, seat belt and harness mounts, tach mount, gas pedal, interior sheet metal, brake line mounting tabs, rear end limiting strap mounts, and tow hook.

The dry sump tank with two or three gallons of oil in it and the weight of the hoses will weigh close to forty pounds, so it should be well mounted. We installed the tank on the right side of the car in what would be the passenger's area. It mounts at the rear, on the cross bar and on the side on an angle iron bracket that runs from the lower part of the roll bar to the angle bar that goes to the dash bar from the bottom of the roll bar at the frame. Small tubes are welded in the main tubes to retain the original strength.

The steering shaft support should be wider than the one we used. It is difficult to put a wrench on the rod end locking nut. Use a piece of channel or cut a piece of rectangular tubing.

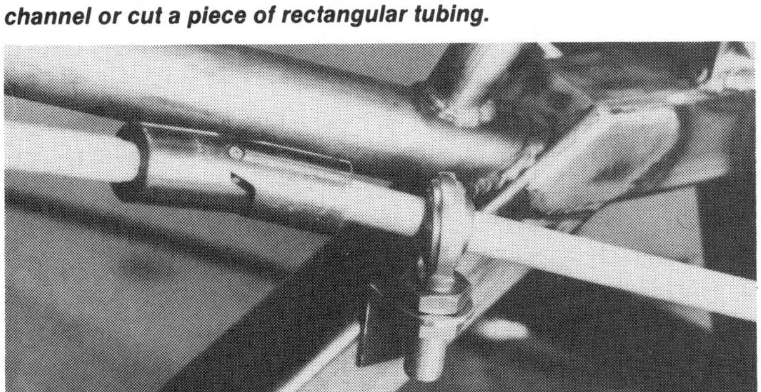

A simple piece of angle iron will make a good steering shaft mount.

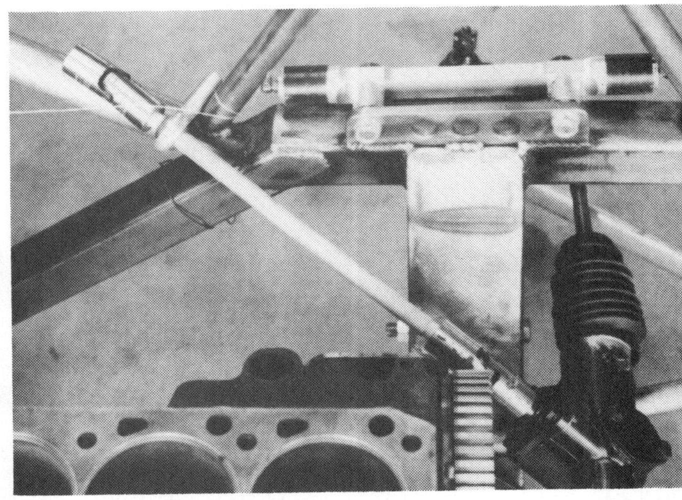

To determine the angles of the steering shaft, use some wood dowels and hold the rod ends with tape.

This type of steering mount is more work and does not offer any advantages

Radiator and cooler mounts should be built to hold the radiators in place but not so solid that they would crack from vibrations or during a crash. Notice the U shaped mounts with rubber padding on Gary Ebeling's car.

These pieces of steel strap were placed here as ballast mounts. In the foreground note the jacking points made of square tubing.

The fuel cell metal box is dropped in a frame made of angle iron and of rectangular tubing at the front.

The fuel pump mount. At the rear note the fuel cell support straps and the triangulated cell protection hoop.

The battery box can be made of 1 x 1 inch angle iron. We fabricated the hold-down bracket, but Mr.Gasket makes one that is universal and easy to install. Weld a battery ground bracket to the frame. Do not run the ground to the battery hold down studs.

The easy way to locate the holes in the pedal mounting bars. The holes in the mounting bars should be sleeved to retain their rigidity. Instead of cutting and welding the bars, have them bent. This will prevent injuries to the driver's legs in case of an accident. Make sure the bars are welded at both ends.

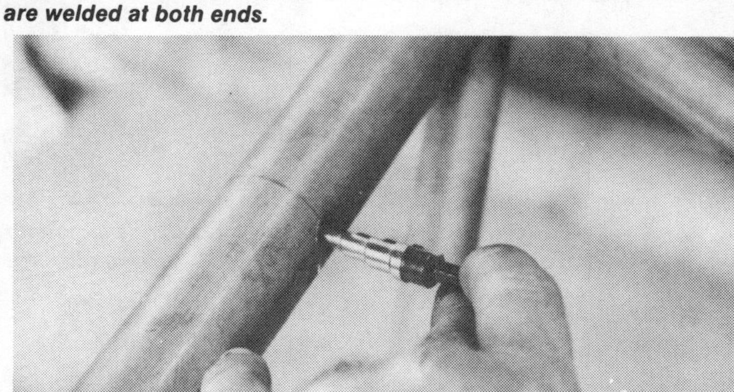

The Cobra bits make perfectly round holes and they are nice to use on tubing. Be sure to use plenty of lubrication when drilling a hole with one of these drills.

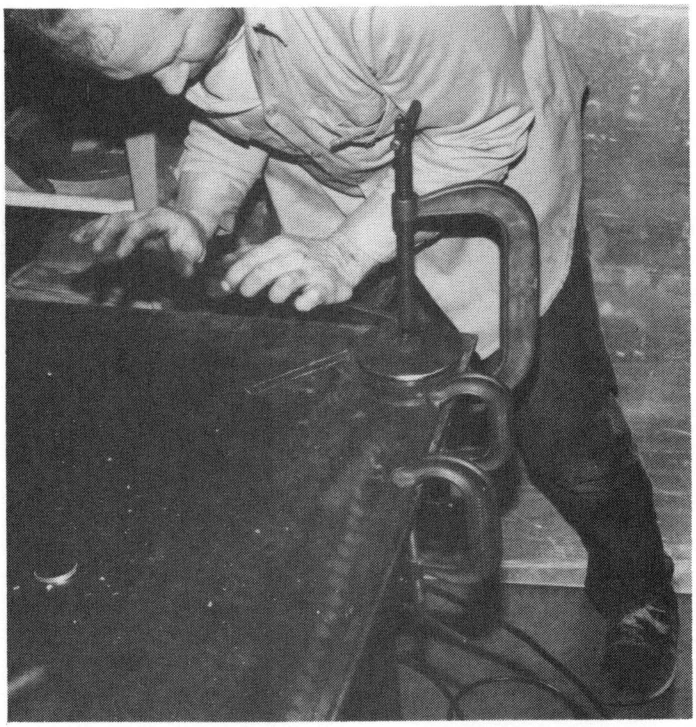

Square tubing bending made easy. Gary Ebeling built this device which is nothing more than a piece of four-inch diameter pipe with a round lid on it. The two c-clamps hold one end of the square tubing and the operator bends the other end around the pipe. Small square tubing is used to hold the body along the lower edges of the car and to make supports for the hood pins.

Dry sump system breather can mount. Note the brake line mounting tab just to the left of the can. The tab is made from a Moroso roll cage gusset.

This tool made drilling holes straight through tubing easy. This tool is inexpensive and it saved us a lot of work. We purchased it at a local hardware store.

Anti roll bar brackets. Run two bolts through a piece of 1/8-inch strap and weld the heads on. Grind the heads off and mount the anti-roll bar brackets on the plates.

We installed this tube behind the driver's seat to run the harnesses at a good angle. Brackets were welded to the bar to anchor the harnesses.

Chapter 5

Interior, Pedals, Gauges

The Throttle Cable

Get the type of cable with the casing made of wound wire material. These cables can be cut and set to length. These can be purchased from BAP auto parts stores. BAP/-GEON stores are Weber dealers and they supply small and foreign car parts.

Once the cable is cut, tap the ends to accept a 1/4-inch lock nut. This will keep the outer cable from slipping out of its stay. Use a sliding barrel and screw on the cable at the pedal to allow the cable to be adjusted to set the pedal height to suit the driver.

You will need to build a plate that mounts on the intake manifold to install the cable at the carburetors.

Pedals

We used the Tilton 72-623 pedal assembly. The Tilton assembly is easy to mount and it is versatile enough to fit any chassis.

Attach the bracing above the assembly. This will keep the mount away from the driver's legs in case of an accident. Also keep the horizontal part of the mount as short as possible so it will not protrude as much near the driver's knees. Mount the pedals so you can use your left foot for braking, if you run oval tracks, and so you can heel and toe for road racing.

The easiest way to mount the pedals is to install two bars from the firewall to the dash tubes. The two tubes should be drilled and sleeved to retain structural rigidity. The sleeving could be done an inch apart to allow adjustability. Use square steel tubing of 1-inch by 1-inch, by .095-inch wall. Be sure that both ends of each tube are welded in place. Never leave one end unsupported or the weld at the other end will break.

Get a pedal assembly for dual cylinder mounting. Using

The interior of our Ford is very functional. The Ford Motorsport gauges are tightly grouped and the tach is out of the driver's view when turning left, but it is easily seen for up-shifting. The brake bias adjuster was installed near the gear shift lever and the electrical switches are at the left behind the quarter window. The Moroso oil pressure warning light is in plain sight and the Tilton remote brake and clutch fluid reservoirs can be easily checked.

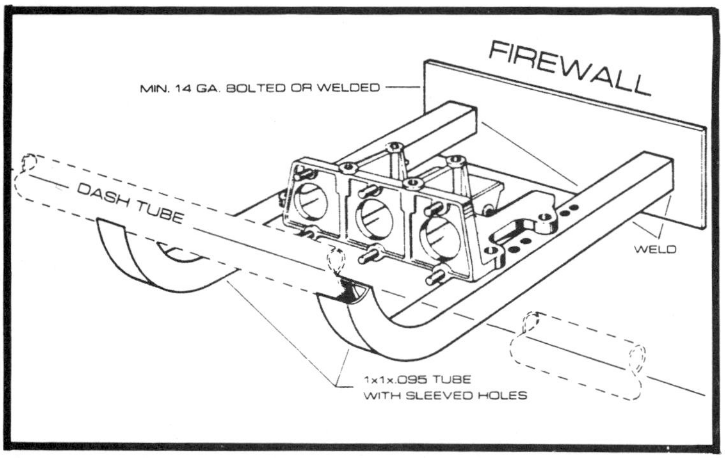

The Tilton dual pedal assembly can be mounted in a number of ways, but on a tube frame such as this is the most advantageous because it allows the pedals to move forward or backward to suit the driver.

two master cylinders, one for the front and one for the rear brakes, allows the use of at least half the system braking ability should the brakes fail at one end of the car. The second advantage to the dual cylinder set-up is that this system allows the use of different size master cylinders to tailor the brake proportioning. The third advantage to the dual master cylinder system is that the car can be equipped with an on-board brake bias adjuster for the driver to fine tune the front-to-rear braking bias. The balance bar remote adjuster is Tilton part number 72-501. Steve Smith Autosports also offers a blueprint for building this pedal mounting assembly and bias adjuster — part number DB3.

The remote adjuster allows the driver to compensate for changing conditions such as fuel loads, track conditions, spring changes, driving style, tire pressures and compounds, etc. On a short track, when the brakes are balanced correctly, the driver should not have to turn the adjuster more than 1/2 turn in either direction to get correct brake balance. 1/4 turn will make a noticeable difference and I found that when the bar is set-up neutral, 1/4 turn to the right will cause a push and 1/4 turn to the left will cause the rear end to come around. The trick is for the driver to dial-in as much rear braking without spinning out. If you are not well experienced, install a bit more front end braking, instead of risking a spin. When installing the remote adjuster, make sure you tighten the short aluminum end that threads into the balance bar. The end has an Allen screw that secures it to the cable and a jam nut should be installed on the balance bar to lock the cable end. If you do not use a remote cable adjuster, make sure the jam nuts are tight to prevent the balance bar from threading itself out.

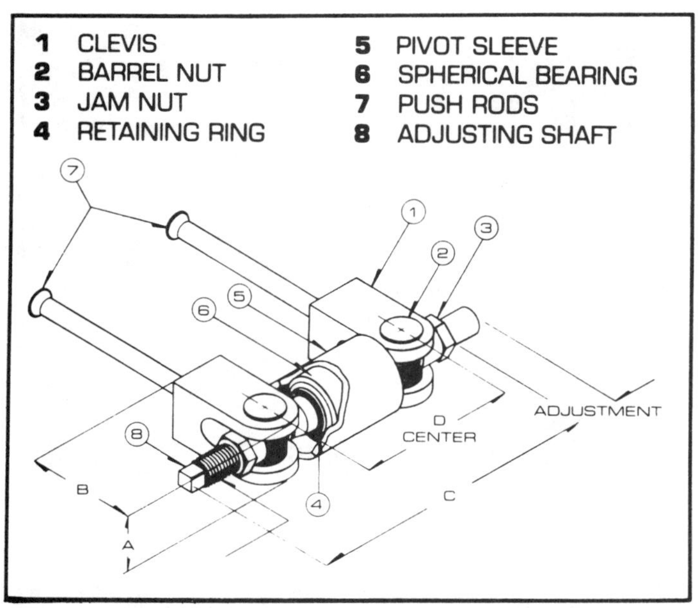

This is the balance bar from Tilton Engineering. A blueprint for building it is also available from Steve Smith Autosports (order #DB6).

After the balance bar is installed, the brakes are bled and the pedal height is set, apply the brakes and set the balance bar parallel to the pedal assembly (perpendicular to the frame). If you run the remote adjuster, remove the jam nuts on the adjusting shaft. If you don't, you will not be able to turn the shaft with the remote adjuster. Another adjustment that must be made on the balance bar is the parallelism of the master cylinder push rods. The center of the clevises at the adjusting shaft must be the same width as on the push rods where they enter the master cylinders. The Tilton catalog contains all the instructions, dimensions and specifications needed for designing the system and its installation.

Brake Proportioning Valves

The brake balance bar balances the braking from front to rear without increasing the pedal effort. The brake proportioning valve reduces the amount of braking at the wheel or wheels the valve is restricting. These valves are used mostly on dirt cars where the driver wants to set his braking according to the track conditions and on road racing cars that carry a large load of fuel, which will be used up during the race.

Some racers use the brake proportioning valve in place of the brake balance bar when a single master cylinder is used. But, for better safety, adjustment and braking control, a double master cylinder set up with a balance bar is the best arrangement.

Clutch and Throttle Pedal

If you want to use a hydraulic clutch, get the three pedal assembly from Tilton. If you want a mechanical linkage clutch, you can also use the Tilton assembly, fabricate a pedal or buy one from Ebeling Engineering. The stock Pinto cable can be used, or Ford Motorsport has a heavy duty assembly, part number M-7553-A302. This kit consists of a beefed-up clutch release fork, and a heavy duty

Teflon-lined clutch cable.

The throttle pedal is easily fabricated using 1/4-inch steel wire and a short tube with a 5/16-inch hole. A bolt is inserted through the tube as a pivot point and a mount is fabricated out of a short piece of channel the same width as the pivot tube. Take care to use the bolt as a pivot point only and not to over tighten the bolt, or the pedal will stick. And, of course, use a locking nut on the pivot bolt. Lube the bolt and use a washer at each end of the pivot tube. At the upper part of the pedal, weld a flat piece of 1-inch steel strap with several holes to allow flexibility of adjustment and pedal ratios for the cable. At the bottom weld a flat piece of sheet metal for the right foot to rest on. As a final check, look for components that could get in the way of the pedal and throttle linkage and could cause a stuck throttle. If you road race, design the pedal for easy heel and toeing.

Warning: Many racing accidents are caused by a stuck throttle. Design and install the pedal with extreme care to prevent this from occuring on your car. Preferably, purchase a pedal from a reputable aftermarket dealer.

Both the clutch and gas pedals need stops. Drill a 5/16-inch hole behind the pedals in the fire wall and weld in a 5/16-inch nut inside the driver's compartment to run the bolt through. Install a 5/16-inch bolt from the inside of the car and run a lock nut from the outside of the fire wall. Because the gas pedal was considerably narrower than the clutch pedal, we welded a washer behind the bolt head to stop the pedal.

Gauges

Use at least a water temperature gauge, and an oil pressure and oil temperature gauge. For road racing, a transmission and rear end temperature gauges would be advisable. A separate volt meter would come in handy to indicate when the battery is fully charged and when it is

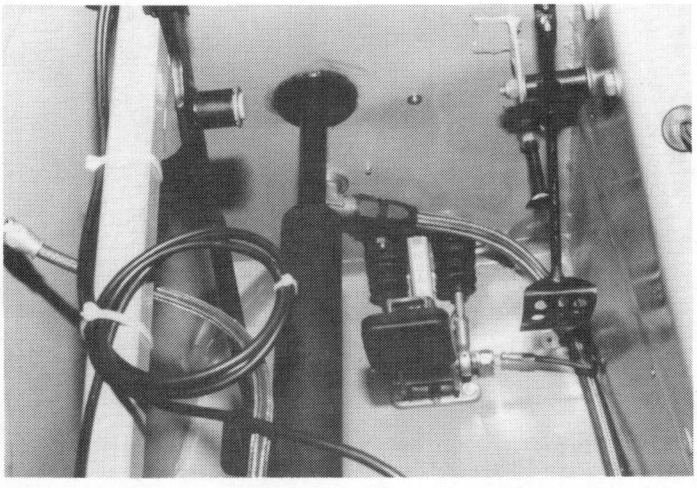

A stock Pinto clutch pedal can be used with the mechanical linkage. Note gas pedal installation, the stop and the padding on the steering shaft. The large braided line is for the fuel and the small one on the left is for the oil pressure gauge.

Mounting the water and oil temperature and oil pressure gauges in this fashion makes their reading very quick. We bolted the panel to the mounts but it could be welded. We used a Grant 13-inch steering wheel. Use a larger wheel as a small diameter makes the steering quite heavy.

We originally used lights in the gauges but they were not needed even at night. A braided line goes into a T fitting behind the oil pressure gauge and an adjustable warning light sending unit fits at the other end of the fitting. Arrows point to the fire bottle knobs. Note their strong mounts.

starting to get low. Ford Motorsport gauges were used on our car.

A fuel pressure gauge is not necessary and it is not advisable to install one inside the car, because of the fire hazard.

We used a Ford Motorsport tachometer (part number M- 17360-B991) because of its memory feature. This tach can recall the highest RPM reached during a race. This is a very important advantage because it allows the driver to concentrate on the driving instead of looking at the tach. The highest reading can be studied to gear the car for maximum performance. The gearing should be set so the engine turns 200 to 600 RPM over the maximum power RPM peak.

Group the gauges so they can be seen at a glance by the driver. We set our gauges in a triangular fashion in the dashboard. Gauges all set in a row require the driver to turn his head and to take his eyes away from the track to be seen. The gauges should be mounted so that the needle of each gauge is pointing straight up in the normal operating mode so a quick glance will tell the driver everything is normal.

As a low oil pressure warning device, we used a Moroso adjustable oil pressure warning light. The pressure sending unit is adjustable from 15 to 50 PSI, to warn the driver when a low oil pressure condition exists. The Moroso adjustable oil pressure warning kit carries part number 49500.

The Steering Wheel

We used a Grant 13-inch wheel to fit the Moroso quick release hub. Use a 15-inch wheel. The smaller steering wheels are more attractive but they are more difficult to steer and to see the gauges through. Before choosing your wheel, make sure you can get a quick release hub that will fit the wheel.

Miscellaneous

The fire bottle knobs should be mounted within easy reach of the driver and at an angle which allows the driver to strike the knobs with enough force to set them off. Remember that what appears easy in the shop may not be so easy upside down, with a concussion or a broken arm. Also mount the knobs on solid brackets, not on sheet metal. And, see to it that they are not in the way of the driver when he enters or exits the car, to prevent accidental set-off.

The electrical panel on our car was mounted in an easy-to-reach area. However maintenance is not so easy. Mount the panel where it is easy to reach and service, and take care that the switches are not too close to the steering wheel to prevent accidental switching while driving.

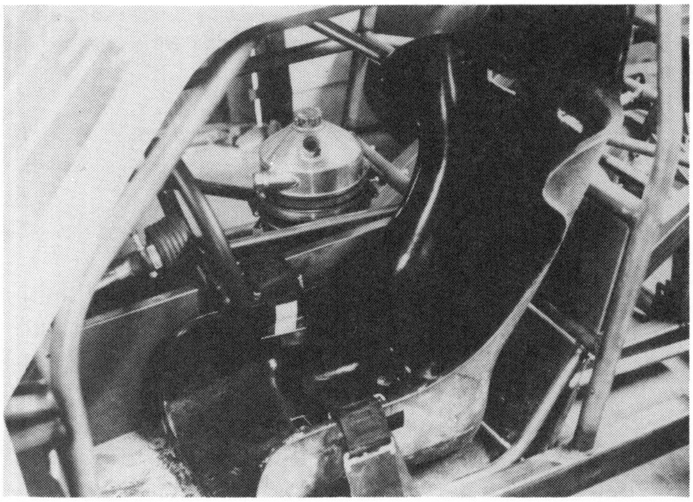

Our seat from Genesis Fiberglass Products gives good support to the shoulders and the hips. This seat also comes with a head rest built in.

The seat should be secured to the floor and to the seat hoop preferably with a full frame. Tubular seat frames are available from Speedway Engineering. This bracket was fabricated to anchor the front of the seat to the floor.

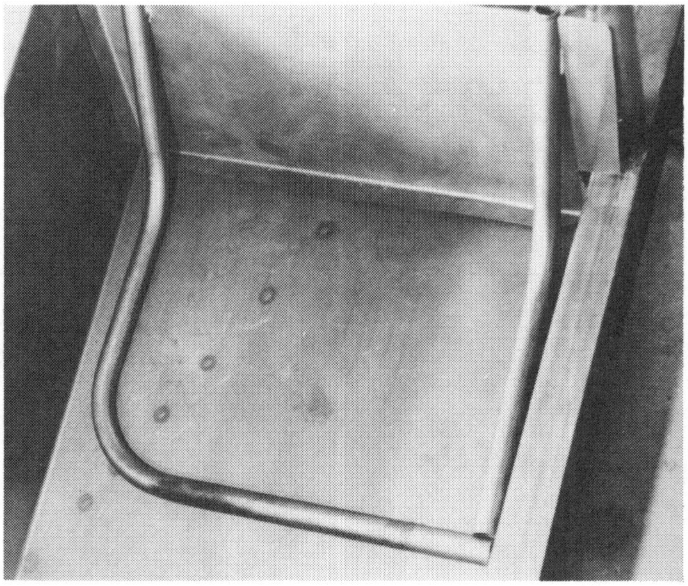

The seat hoop is an important safety feature. The seat is fastened to the hoop and the hoop is welded to the frame. Should the car get hit in the left side, the seat will move to the right with the hoop. Just be sure the lap belts are also secured to the hoop.

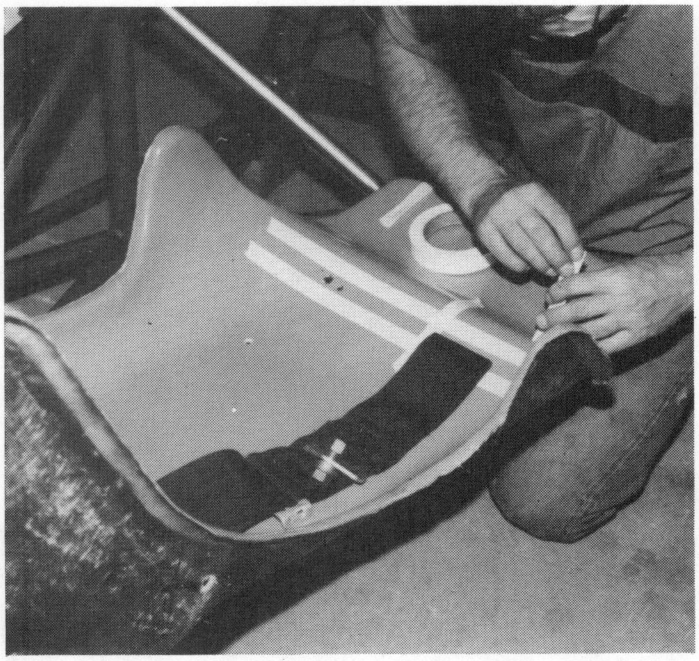

Drill a small hole in the seat and run strings to find the location of the openings for the harnesses. Using masking tape like this makes an easier and neater job.

This is how much sheet metal a standard chassis comes equipped with. The main object of building the interior sheet metal is to provide a fire proof barrier for the driver. Fit the sheet metal as tight as possible to keep dirt out of the cockpit.

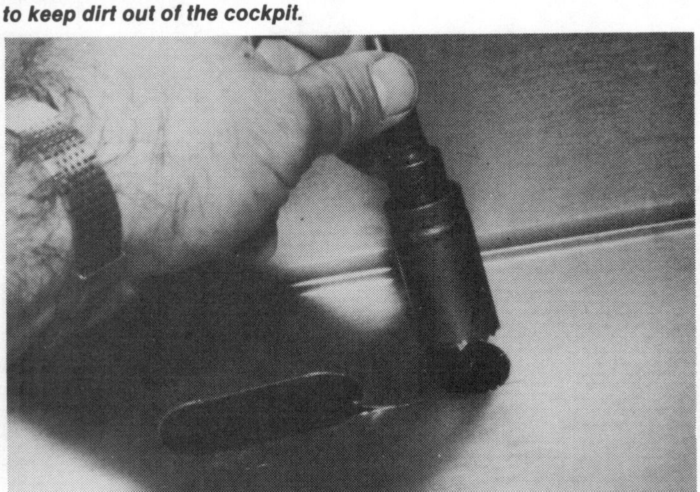

Round holes can be cut in sheet metal with hole saws. Body punches make a neater job but they are expensive. A slot like the one on the left can be cut by cutting two holes with the hole saw and sawing the metal in between with a fine tooth jig saw.

Interior sheet metal should be made out of 2024 or 3003 T3 aluminum, 0.030 or 0.040-inch. This piece which we used at the front vertical section of the tunnel, just to the right and forward of the gear lever, shows all the operations that need to be performed to fabricate the interior sheet metal: cutting, drilling, bending, slotting. Once the piece is finished, its edges should be filed and sanded to remove the sharp edges.

Make a pattern out of cardboard and lay it out on the aluminum after it has been fitted properly.

The tunnel can go straight across the driver's compartment or it can come across to the right side window. We built our tunnel up to allow more air flow out of the radiator and to the back of the car.

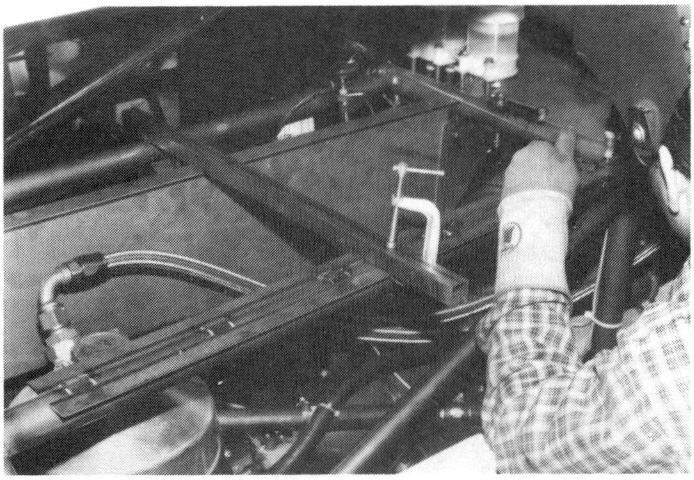

Steel strap 1-inch wide by 3/16-inch thick must be welded in place before the sheet metal is installed. The piece of square tubing with the clamp is used to hold the piece of strap flush with the top of the cage tube while it is being welded.

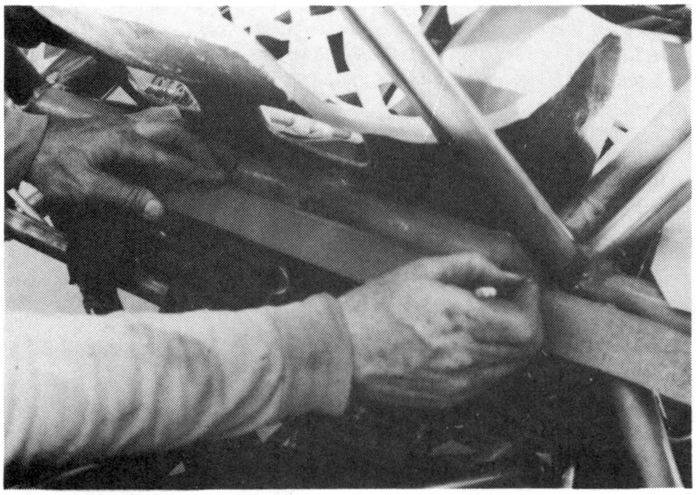

Steel sheet metal can also be bent and welded to the chassis to fasten the sheet metal to.

This opening was cut to allow servicing the oil tank.

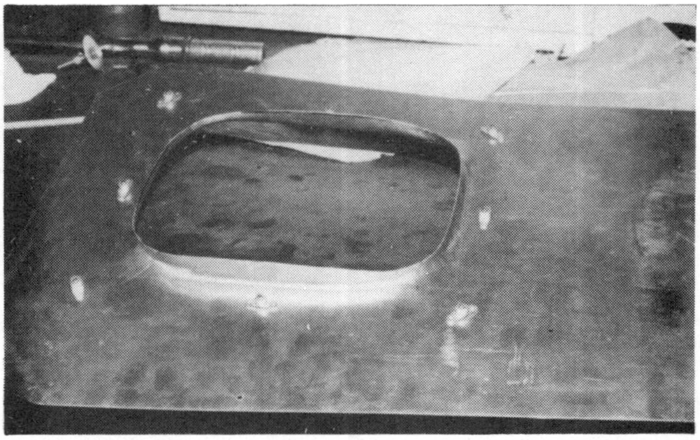

The hole for the oil tank is strengthened by shaping a lip around it. This lip is built with a plastic hammer, using a wood block as a back-up. Nut plates are riveted in place to hold the cover on.

Rolling a bead on the sheet metal will greatly increase its rigidity and appearance.

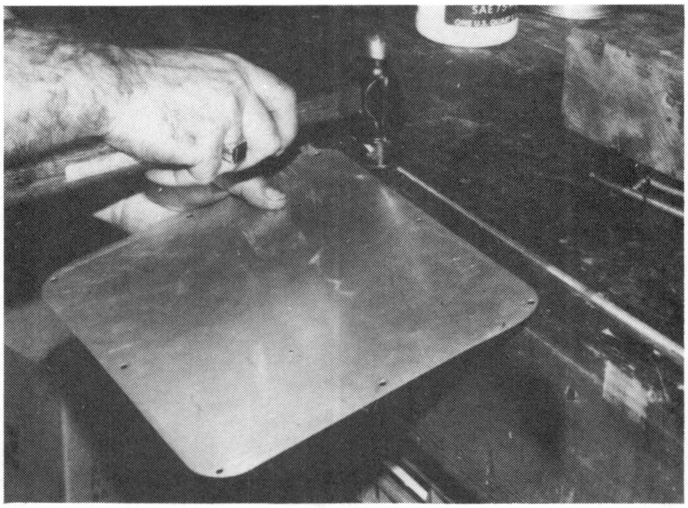

Pounding the edges of a flat piece like the cover to the oil tank opening stiffens the part and it gives it better appearance. The lip will also seal the outer edge of the cover to make it air tight.

The body must be off the car to install the sheet metal, but the body mounts to the sheet metal in several places and the aluminum must be fitted to the contour of the fiberglass. Here the rear deck is trimmed to fit around the roll bar rear supports.

The rear window shelf and the body are riveted together at the rear and one of the sides. A second piece will be added to finish off the right side. The rear shelf also rivets to the front to the rear deck.

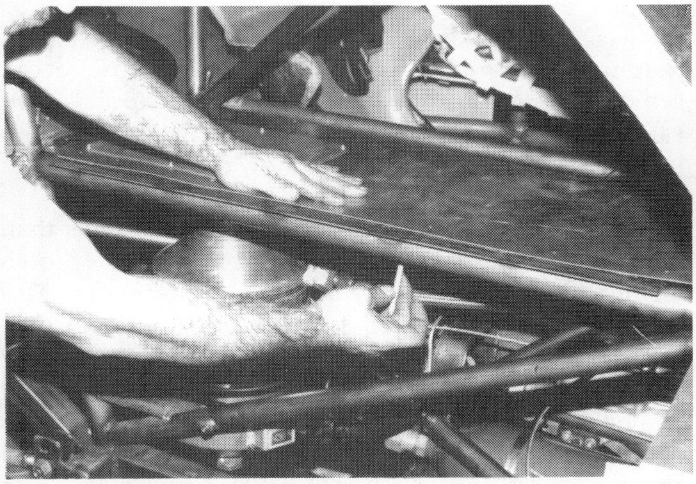

The more simple pieces can be made without a pattern. Lay the piece on top of where it will be mounted and trace the outline with a pencil. By measuring half the width of the steel strap that was welded to the chassis outward of the line you will end up with the location for the line the rivet holes will be drilled on.

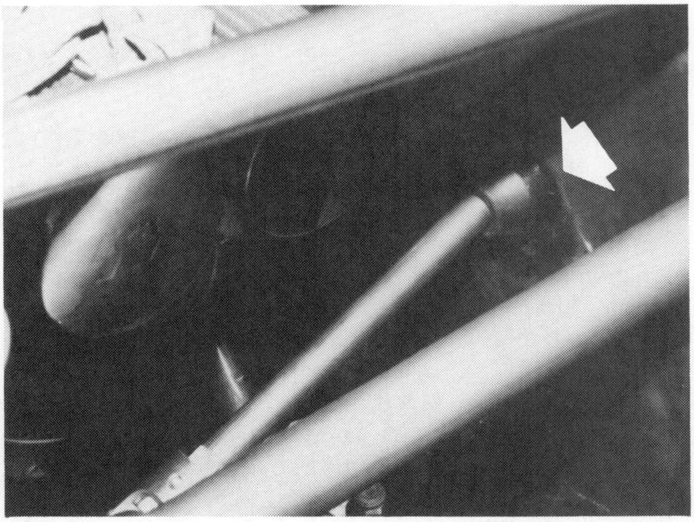

A ball joint boot makes a good steering shaft-to-firewall insulator.

Chapter 6
The Front Suspension

The suspension design on the modern day tube chassis Mini Stocker is very straight forward. The general layout is comparable to larger race car suspension systems, only down-sized to go along with the lighter weight and shorter wheelbase.

The geometry involved in the design of a Mini Stock front end is basically the same as any other racing car. However a few factors must be kept in mind: 1) the racing weight of a Mini or GT car is much lower than the full size Modifieds or V8 SCCA or IMSA road racing cars. 2) The center of gravity height in a Mini is very low in comparison with the bigger V8-powered cars. This consideration is particularly evident with a 2300 cc car, where some 600 to 700 pounds of weight have been added low in the chassis. 3) The power output of a four cylinder, normally aspirated motor is much lower than a V8. 4) The wheelbase is shorter than the bigger cars and the track is narrower.

Nonetheless, off-the-shelf parts such as springs, shocks and anti-roll bars can easily be purchased for the Mini or small GT car from suppliers such as Carrera Shocks, Ebeling Engineering and Speedway Engineering.

Basic Design Principles
Bump Steer

Bump steer is, simply put, the introduction of toe (either in or out) into the front wheels when the suspension goes into bounce or rebound. Bump steer can also occur in the rear on a fully independent suspension and a related phenomena, called roll steer, occurs in solid rear ends as the chassis rolls while negotiating a turn. If bump is not set, a bumpy track will cause the wheels to travel up and down causing a change in toe, while braking will cause both front wheels to bump again, possibly changing the toe at one of the most critical times.

On a short oval track, the speeds are usually lower than on a road course but braking is done with greater force mainly due to the left-turn-only set-up of the cars. This will cause more front weight transfer and more wheel bump. Lastly, toe may also change if the bump is not set properly, during cornering due to body roll. This should be checked while checking bump steer by turning your front wheels say half a turn of the steering wheel and checking the bump in one inch of travel up on one side and an inch of travel down on the other side. Hopefully one side will toe out as much as the other side will toe in.

Bump steer occurs when the tie rod end follows a path different than the path the wheel is following. If you visualize the arc created by the front wheel as it moves up and down through its travel, along with the arc created by the tie rod end , you will see that both arcs must have the same instant center or the tie rod end will move in or out in relation to the wheel, causing a change in toe. If the arc created by the tie rod end and the wheel are parallel, no toe change will be present.

The easiest way to design a no bump steer system in a stocker or sedan is to place the tie rod end directly in line with the ball joints and the inner tie rod locating point, at

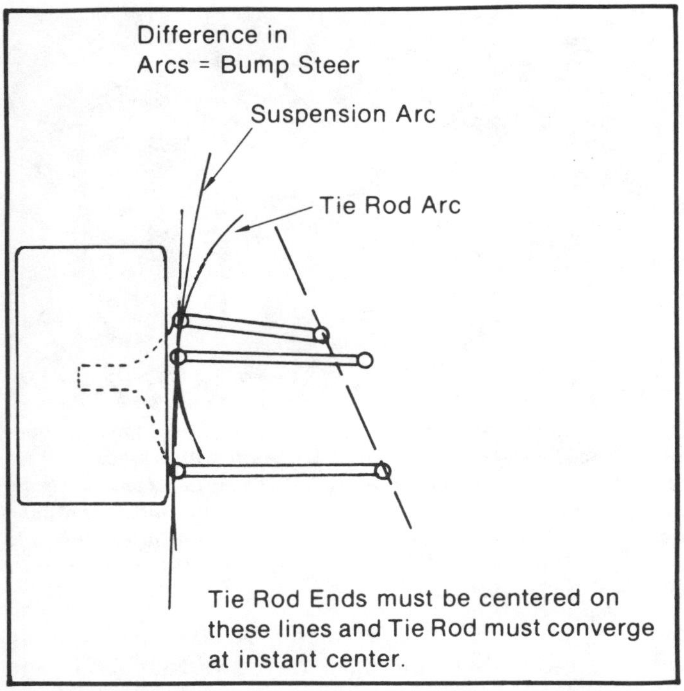

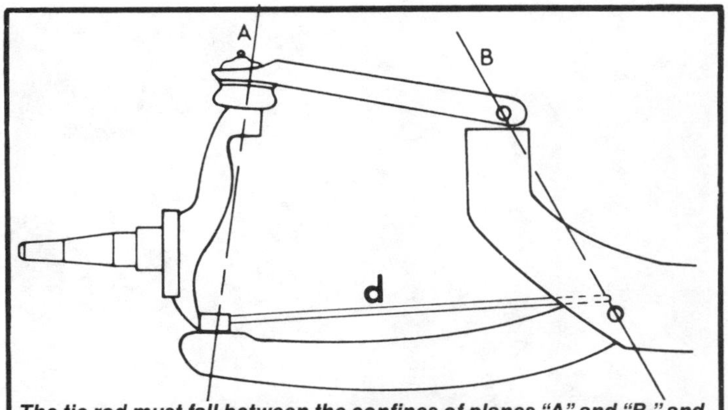

The tie rod must fall between the confines of planes "A" and "B," and the tie rod's centerline must intersect the instant center. The tie rod may be laterally displaced so long as the length remains the same as if it were between "A" and "B," and the centerline still intersects the instant center. This will yield zero bump steer.

the rack, should be directly in front of the lower control arm locating bolt. In other words, the lower control arm mounting bolt, if it were extended out, should cross the center line of the inner tie rod end.

Steering Rack Centering

It is very important, when adjusting the toe to have the steering rack centered so the tie rod inner mountings are directly in line with the lower control arm locating bolts when the car is driven in a straight line. To find the true rack center line, not the housing center, install the rack in the car and align the tie rod inner joints with the lower control arm locating holes. Measure the distance between the tie rod lock nut and the housing for future reference or, better yet, make a tool the same thickness.

The next time you want to set the toe, simply slip the boot off the end of the rack on the same side you mea-

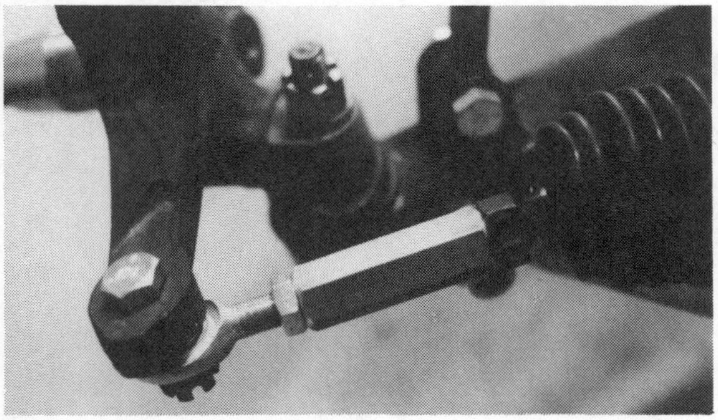

Steering shaft extensions for use with a rod end can be made with 3/4 or 1-inch hexagon 2024 T6 aluminum.

Once you have found the center of the steering rack measure the distance between the rack and one of the knuckles. With this figure you will be able to center the front end before setting the alignment. We made this tool out of some scrap metal. The tool can be inserted between the knuckle and the rack for a quick and accurate method of centering the front end.

sured and insert the tool to center the rack. Now each front wheel can be set individually. This will also keep the steering wheel centered so you can still see the gauges.

Another factor affecting bump steer is steering rack location fore and aft. With our car we found that locating the rack about an inch ahead took all but about .010-inch of bump steer out. The rack needed to be moved forward to fit the dry sump drive pulley any way.

Setting Bump Steer

The bump steer is most easily checked using a gauge manufactured by Vic Irvan Enterprises. The major difficulty in checking bump steer is making a stand that is stiff ang strong enough to give accurate readings. The Irvan gauge is reliable, it provides accuracy and it is fairly inex-

The Speedway Engineering rack and pinion will also fit a mini stock chassis. These units are very sturdy and give better steering feedback. The racks can be made in any length desired. Note sturdy u-joint.

The stock Pinto steering rack is equipped with three mounts. The mount near the center of the rack can be cut off. Moroso makes a solid mounting kit for the Pinto steering rack. This kit can be used in road racing, but on a short track the rubber mounts will take up some of the shock from the contact with the other cars.

The steering arms on the Pinto racks are held in the socket by a large jam nut. The pins should be replaced each time the steering arm is removed.

Steering u-joints should be of the highest quality. Use joints that have a solid block between the two halves. Some joints have blocks that are split in half and they are joined with a bolt. These joints are not strong enough and should not be used on race cars. To prevent play between the shaft and the joint, install a heavy roll pin through the joint tube and the shaft and then use a 1/4-inch grade 8 bolt through the roll pin. Note safety wire on the Allen bolt.

pensive. to set the bump steer first set the alignment within a reasonable limit, place the car on jack stands and disconnect your anti roll bar, springs and shocks. Remove wheels and bolt the hub plate provided by Irvan or your home-made plates if you built your own stand. Now place a jack under the lower A-frame and level hub plate.At this stage I found it easier to set the hub at ride height and check the toe change through two inches of bump and then go back down to check the toe change through an inch of rebound. This way I had a reading below and above ride height, with ride height being zero and set them level on the plate. Jack in one half inch of travel at the time and note the the dial indicator readings. For circle

The Irvan bump steer gauge consists of a frame supporting two dial indicators and a hydraulic jack. A graduated plate bolts to the hub. As the jack moves up the plate moves across the dial indicators to show bump steer.

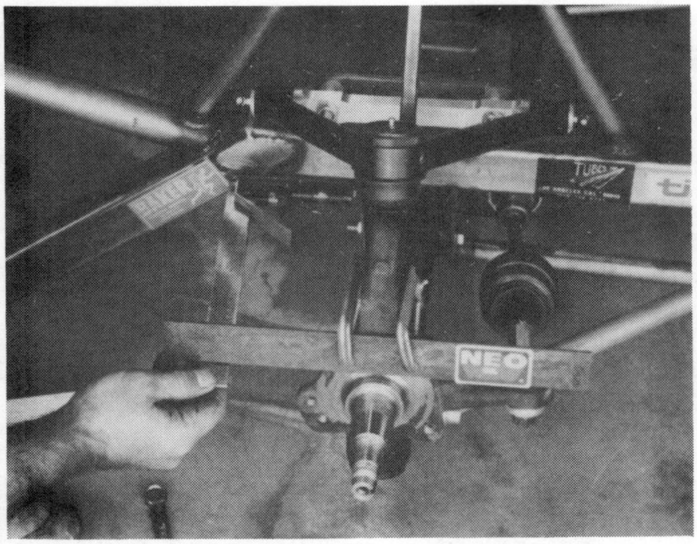

Setting the bumpsteer can be done using this method. This is not as accurate as using the Irvan bump steer gauge.

Steering arm on the spindle on the right was lowered to set bump steer. If the rules allow the engine to be set back, you will be able to raise the steering rack enough to line it up with the steering arms. The spindle on the right was also lightened.

track racing as much as .050-inch is acceptable while for road racing as close to zero as possible is preferable.

For dirt racing, bump steer is actually recommended. Irvan recommends the left front wheel to bump out .015-inch and the right front .030-. The reason for this departure is that in the dirt, the steering is cranked to the right. By increasing toe out on the right side during bump and decreasing toe out on the left during rebound we retain the toe out that was originally dialed into the front wheels. Introducing some bump steer in the front end for dirt racing reduces push going in and reduces oversteer coming out of the turns.

To change bump steer with the Pinto/Mustang spindle it is easier to use the tie rod end adaptors sold by Baker Precision. This adaptor slips in the hole in the spindle arm at one end while the other end is straight allowing a rod end to be used. The adaptor allows placement of washers over the rod end to drop the tie rod. Using the stock spindles and tie rod ends without the adaptors, the spindle arms must be heated and bent to reset bump steer (which is definitely not recommended). The problem with this method is that repeated heat cycles will embrittle the metal and may cause an eventual failure.

Another way of changing bump steer is to raise and lower the steering rack or move it fore and aft. I found the stock set up on the Ebeling chassis to be just about perfect for circle track and road racing.

Roll Centers

It was mentioned earlier that the center of gravity height is lower on a Mini stocker due to the extra weight (ballast) that must be added in a tube chassis car. It was also said that this fact will have to be kept in mind while preparing

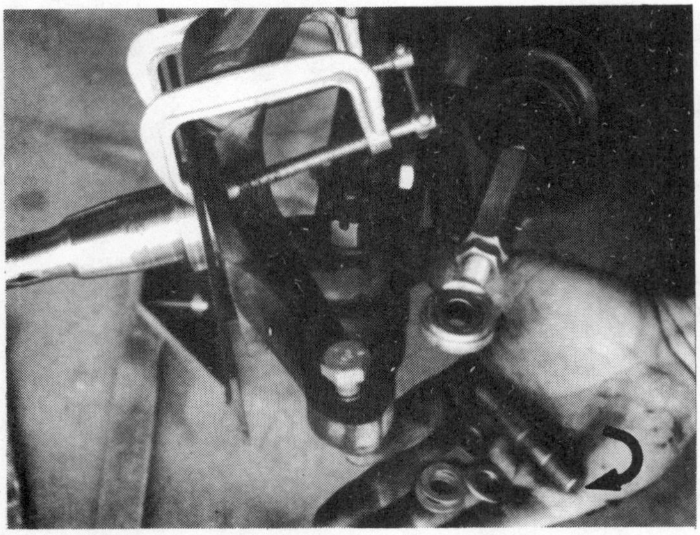

These studs allow the replacement of the stock tie rod end for a spherical rod end bearing. This facilitates setting the bump steer and improves steering feel. The studs are available from Baker Precision.

When installing the lower control arm in the chassis, leave the bolt loose until the car is set down to approximate ride height. If the bushing is tightened when the control arm is hanging down, it will twist and deform when the car is set down. This could damage the rubber bushing and cause handling problems. Also install the bushing so the split in the sleeve aims toward the inside of the car. This way, if the wheel gets hit, the sleeve will press against the rubber without splitting the sleeve.

Locating Roll Centers

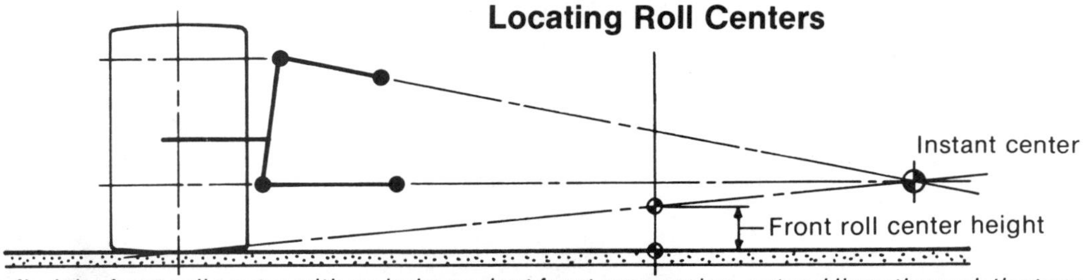

To find the front roll center with an independent front suspension, extend lines through the two pivot points of each A-arm until they intersect. This intersection point is called the "instant center." A third line is extended from the instant center to the tire centerline. Where this third line crosses the vehicle centerline is the roll center height.

When the rear suspension is equipped with coil springs or coil-over units, the rear roll center will be located at the vehicle centerline at the height where the Panhard bar or other lateral locating device crosses the centerline.

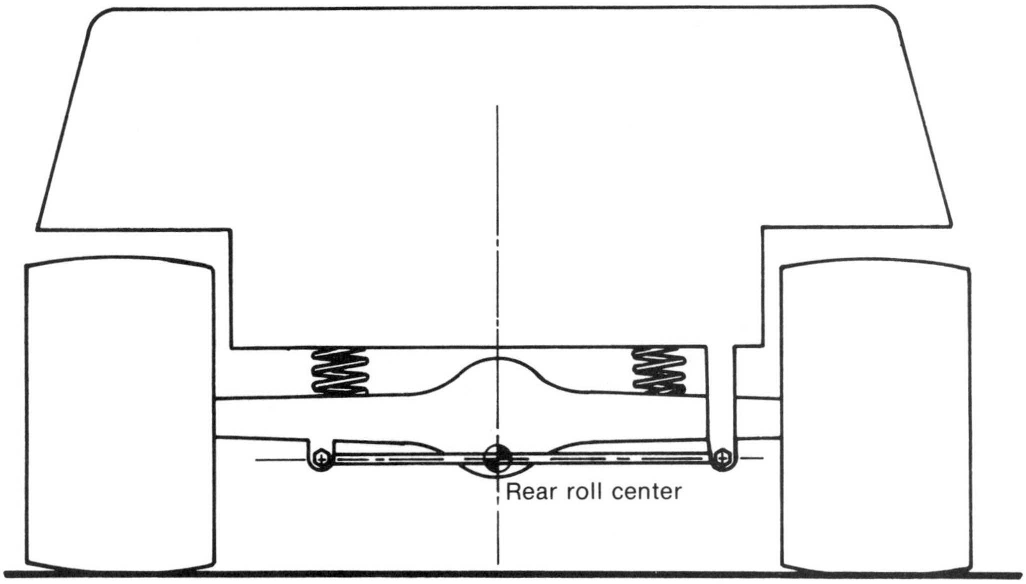

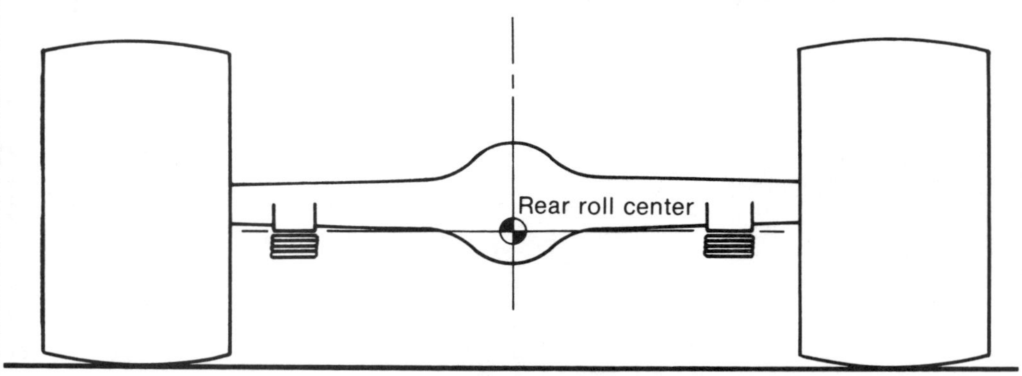

When a car is equipped with a solid rear axle housing and leaf springs, the roll center is located at the vehicle centerline at the height where the springs anchor to the axle housing tube. To lower the roll center, a lowering block must be placed between the housing and the springs. In this case, the roll center will fall halfway between the springs and the axle housing.

our Mini stocker. The location of the roll centers should be determined in conjunction with the center of gravity height because the center of gravity acts as a lever arm against the roll axis to allow the chassis to roll.

The stock Pinto/Mustang spindle will usually yield a front roll center height of about 4 to 5 inches. With an 1800 pound car and a 12-inch CGH, this is an acceptable situation. But when 500 pounds are added to meet the 2300 pound class requirements of most Mini Stock associations, the CGH is lowered to 10.5 inches (when the ballast is added at frame level).

This shortens the lever arm between the CGH and the roll axis, and the car will tend to skate sideways and push in the corners. This means that with the heavier weight and the lower CGH, the front roll center should be lowered for proper handling.

Ebeling Engineering has available a spindle that lowers the front roll center two inches. This spindle would be

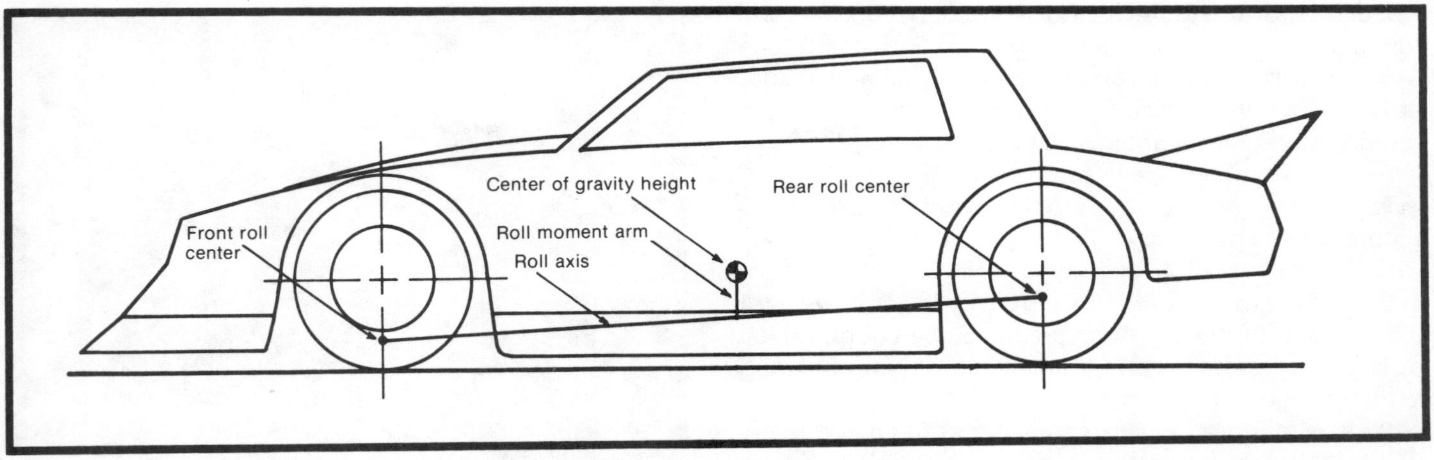

desirable for the heavier weight cars or for drivers wishing to run road racing and Mini Stock races with the same car.

Ackermann Steering Principle

The Ackermann steering theory states that a line drawn through the centers of the front spindle steering arms should converge at the instantaneous center at the center of the rear end housing. The idea behind this principle is to turn the inside front wheel more than the outside front wheel during cornering. The reason for doing this is that the inside wheels are turning around a tighter radius than the outside wheels are.

In the past the generally accepted practice had been to minimize the Ackermann effect on racing cars, but recently some small formula car builders have experimented with more Ackermann with great success. The major reason for this marked improvement is that, without Ackermann, the inside front tire is doing less of the work, thus running cooler, and the outside has to do more of the work because the driver is cranking the wheel more. When the inside front is aimed in the right direction it will stick better and the outside will need less steering angle. Toe out is still important to use to help the car turn at the onset of the corner.

Ackermann steering also offers the nice advantage of increasing at a rising rate— the more steering that is fed in the front end, the more the inside front wheel will turn in relation to the outside. In other words, if the right front wheel angles in 5 degrees, the left may angle out 5.5 degrees, while if we turn the right front 10 degrees, the left may turn 11 or 12 degrees.

This can be observed on a passenger car when the front

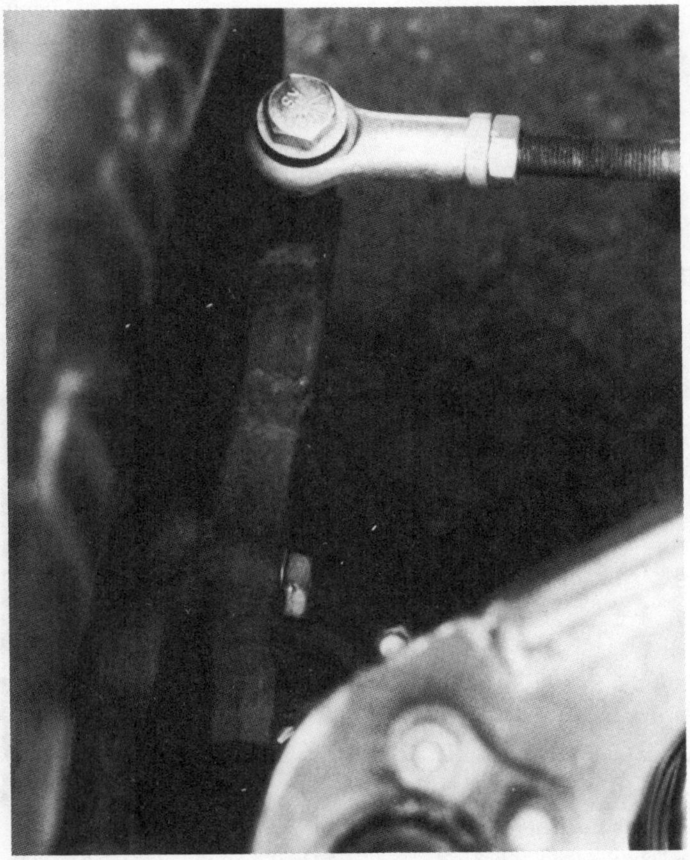

This spindle has no Ackermann built into it. This is evident by the way the steering arm aims straight back.

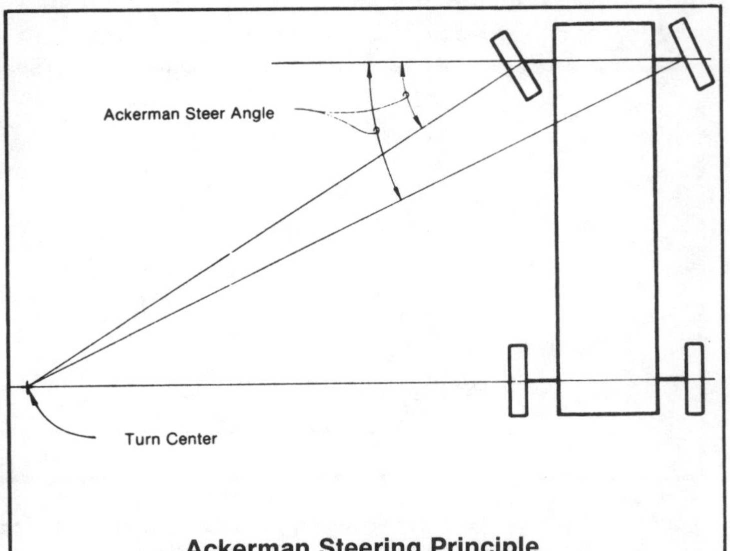

Ackerman Steering Principle

wheels are turned to full lock. The inside front wheel will squeal and it will even leave a tire mark in its path. The Ackermann principle is working nowadays mainly because of the sticky tires which run at low slip angles and that require more gentle driving techniques than the tires of five or ten years ago.

Dirt cars do not need any Ackermann. The reason is simple: when entering a turn in the dirt, the car is sideways with its front wheels turned in opposite lock. If the front end had Ackermann built in, it would over-angle the right front wheel. In the dirt Ackermann can be counteracted by building in some toe-in bump steer in the right front, and/or rebound toe-in at the left front.

In our Mini stock, we can increase Ackerman by moving the steering rack forward in small increments. Remember that too much Ackermann will defeat the purpose and it will induce understeer instead of reducing it.

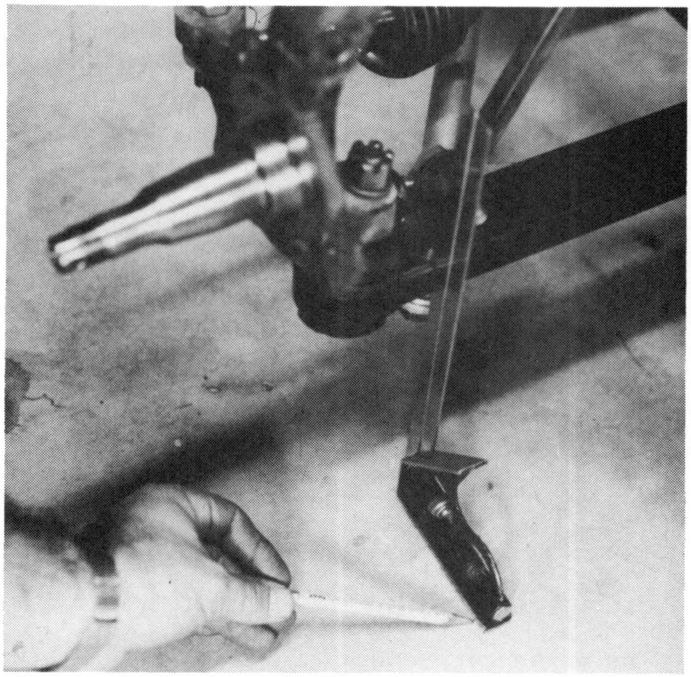

To check caster change in your car, lower the suspension to a couple of inches below ride height and set a square as pictured here against the lower control arm. Make a mark at the end of the square on the floor and raise the suspension to full bump. Now see if the square has moved in relation to the mark on the floor.

Caster

Caster can be visualized when looking at the front end of a bicycle from the side. The rake visible in the forks is nothing more than the caster. Caster in a car is determined by the imaginary line drawn through the upper ball joint down through the lower ball joint and viewed from the side.

Caster is discussed in terms of degrees positive when the upper ball joint is further back than the lower ball joint (the line is angling towards the back of the car) or in terms of negative caster (just the opposite). The effects caster has on handling will be discussed later.

Camber

Camber is the inclination of the wheels at the top, as seen from the front view of the car. Camber is the single most important alignment element in the handling of a race, car and the idea behind setting camber is to maintain a flat tire contact surface with the track. Negative camber is set statically at the right front to counteract for roll-induced positive camber in circle track cars. Body roll causes the outside wheels to move up inside the body and due to its basic design, the unequal length control arms will cause the camber to change towards positive.

At maximum cornering, the tire should be at 1/2 to 1 degree negative camber on the right front and 1/2 to 1 degree positive on the left front for a short track car. To determine your static camber setting jack up a front wheel to its maximum bump travel while cornering and set the camber at 1/2 to 1 degree. Return the car to ride height and note camber for future use.

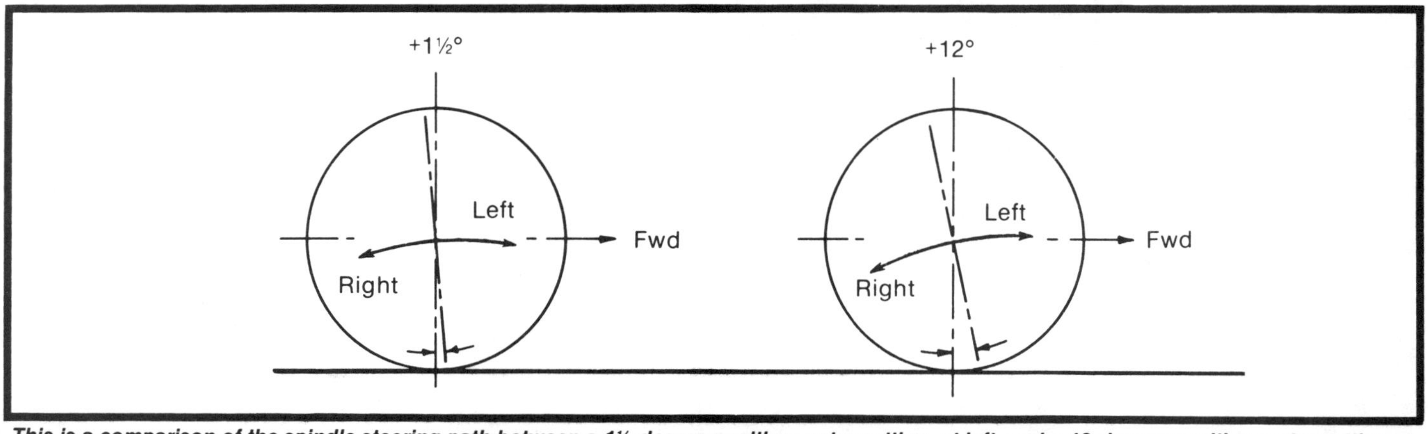

This is a comparison of the spindle steering path between a 1½-degree positive caster setting, at left, and a 12-degree positive caster setting at right. The larger amount of positive caster significantly increases the canted steering path as the wheel is steered to the right. When the steering curve dips below the center straight-ahead height, that corner of the car rises, which means weight is being transferred to the diagonally opposite corner of the car.

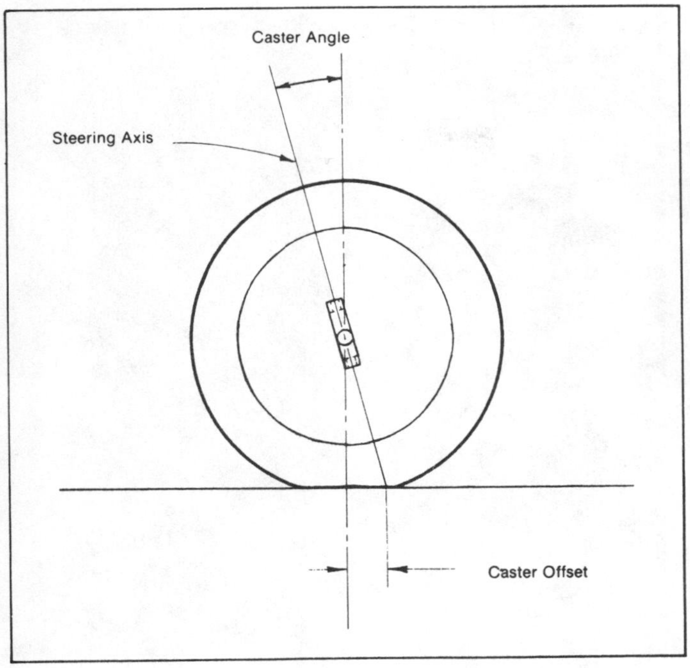

The centerline of the lower control arm bushing should pass through the centerline of the front radius rod rod end ball, otherwise the caster will change as the wheels bump or rebound. This could cause problems if one wheel is allowed to bump, changing the caster setting in that wheel only. Also when the car corners, one wheel will gain caster while the other will lose caster, which could cause some transient handling problems.

The Ebeling chassis is designed with 3/4 degrees of positive camber gain per inch of bump travel. This means that for every inch the wheel moves up into the body, the wheel will gain 3/4 of a degree positive. So for 1 1/2 inches of bump travel, our right front wheel will change 1.125 degrees positive.

To counter this change during cornering, we need to build in 1.125 degrees of negative camber into the right front, minus the half degree negative built up by the caster. So we end up with .625 degrees of static setting plus the 1/2 degree of negative camber the tires like to run with, and we use a static setting of 1 1/8 to 1 5/8 degrees of negative camber on the right front.

A smaller figure can be used on the left front, but it would be a positive setting rather than negative. The left side needs less positive camber than the right needs negative because the left front drops less than the right bumps due to the resistance of the anti roll-bar, and, since the left front is less loaded than the right it will have less tire roll to overcome. The left should be set at around 1 degree positive.

Only skid pad or on-the-track testing with a pyrometer will tell you the correct setting for your particular chassis.

Roll Centers and Anti-Roll Bars

The trend today is toward soft suspensions, low roll centers and low profile tires—the three go hand-in-hand. The lower profile tire reduces tire roll, but the low profile tire also requires more gentle loading and a gentle ride over bumps. The softer springs probably could not keep the chassis off the ground in a turn, so anti- roll bars are used to control sway and balance out the front-to-rear weight transfer under cornering loads.

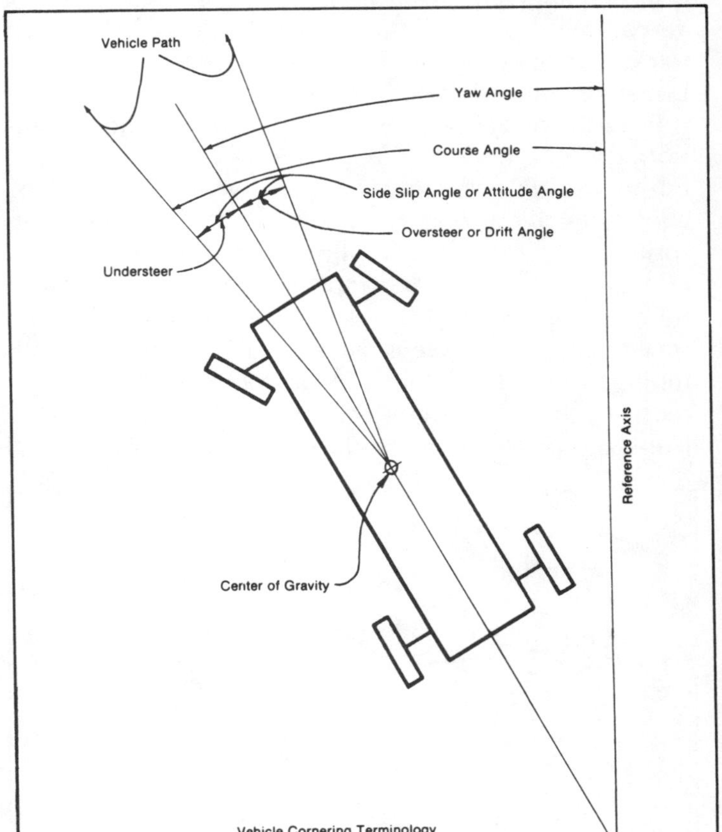

This diagram shows the relationship between a vehicle and cornering attitudes.

The purpose of the anti-roll bar is: 1) to control body roll, 2) adjust roll couple distribution in a simple and easy way and, 3) to allow the driver to change his car handling during a long race to make up for changing conditions or fuel load changes.

The trend toward lower roll centers and softer springs induces a larger amount of body roll which must be controlled somehow.

If the same springs are kept to retain ride height, ball park roll couple, and front-to-rear weight transfer considerations under braking loads, the roll bar should be thought of as a "load transfer agent". This load is still transferred from side-to-side during cornering but when the car is accelearting or braking in a turn, the bar acts as a weight jack and it transfers weight from one corner to the diagonally opposed corner.

An anti-roll bar has no effect on front-to-rear weight transfer under straight line acceleration or braking. But when a car comes off a turn and the throttle is applied, the front bar jacks weight to the inside rear tire (a thought to keep in mind for cars running under rules that do not allow locked rear ends).

One area where anti-roll bar rates become an exception is on banked tracks. The springs on a car run on a banked track must do most of the work due to the downforce while the anti-roll bars take care of the residual body roll. To compute spring rates and roll bar rates on a banked track, read Steve Smith's Advanced Race Car Suspension Development and its companion workbook.

The anti-roll bar can be used for fine tuning the suspension, even from inside the car during a race if you have on-board adjustable bars, so the bar arms should be chosen to allow plenty of adjustability and the front springs should be picked to allow a bar near the mid range rate. This way a softer and a stiffer bar will be available to tune to varying track conditions. Remember that the smaller cars are very sensitive to small changes in chassis tuning. The anti roll-bar can be used, for instance, to reduce a corner exit oversteer condition by increasing the front bar rate and reducing the front wheel rates an equal amount. Adjustable bars are usually used on road racing cars, or oval track cars racing in longer lap events where fuel loads and track conditions change.

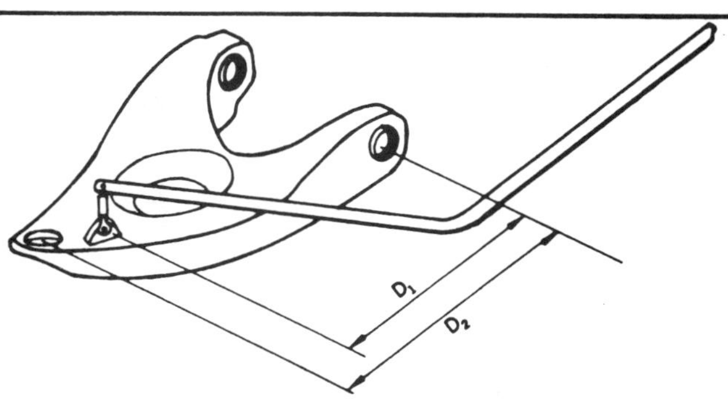

This is how the motion ratio of an anti-roll bar is computed (D squared divided by D² squared).

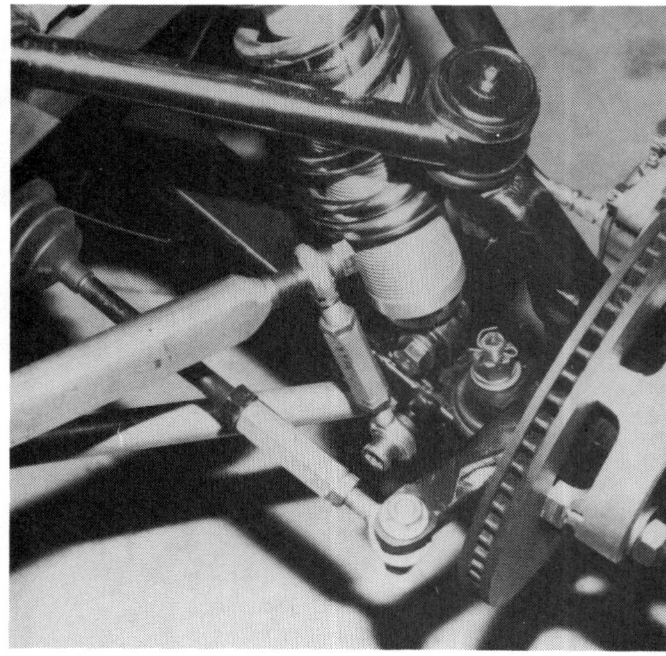

We tried mounting the anti-roll bar attachment point close to the lower control arm ball joint, as sound theory dictates, but this required a radical bend in the arm to clear the tire when the wheels were turned. The bend in the arm and the angle in the adjusters was causing it to twist.

Roll bars for Mini Stocks and road racing sedans are available from Speedway Engineering in two different forms —solid and tubular. The tubular bar is more expensive initially but it has the advantage of being lighter. And, tubular bars no matter which rate bar is used. This last benefit makes the tubular bars less expensive in the long run if you want to keep different rates on hand. The interchangeability with the arms and the brackets also makes the tubular bars faster to change.

Tubular bars come in 1/8, 3/16, 1/4-inch wall, 1.25 inch OD, and in lengths from 32 to 43 inches. We used the 1/8 inch wall by 36.5-inch bar on our project car.

The solid bars come from Speedway Engineering in lengths from 32 to 43 inches and in ODs of 1, 1 1/8 and 1 1/4 inches. The solid bars can also be turned down for special rates. When calculating bar rates, remember to measure the effective length between the two arms rather than the overall length.

Springs

The primary purpose of the suspension springs is to keep the chassis off the track, at a pre-determined ride height and to prevent the suspension from bottoming-out during wheel bump. The springs are also used to balance out the car handling by acting as levers or wedges from

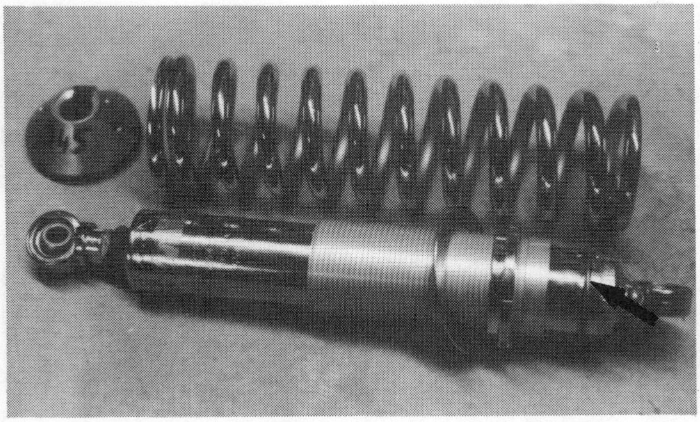

Carrera shocks come with two adjustment grooves to allow the coil-over sleeve to be set in the desired position. We set the front shocks on the lower groove and the rear shocks on the higher groove. Carrera shocks do not use a lock nut on the spring adjusters, so the washer must be used under the springs to prevent the adjustment from backing out. And be sure the car is setting on the springs when running the motor or the vibrations will cause the adjuster to back off. Arrow points to coil-over sleeve adjustment groove.

one end of the car to the other during cornering.

The effect of the springs on the chassis can best be understood if you picture the chassis as a table with adjustable legs. If all four legs are the same length, the load will be equally distributed (assuming the load is placed at the center of the table). If a leg is shortened, the amount of weight placed on that leg will diminish proportionally with the diagonally opposed leg. The weight on the other two legs will then increase by the same amount the shorter legs lost.

If you picture the short leg as a softer spring compared to the the longer legs, you can visualize the effect that say a soft right front spring would have in a left turn— the weight would transfer to the right front unloading the left rear. By the same token if a stiffer spring is used on the right rear it is easy to see that the same effect will take place and weight will also transfer to the right front, unloading the left rear.

The relative balance between the front and rear springs is called the relative roll rate or roll couple distribution. This is better explained in Steve Smith's Advanced Race Car Suspension Development book and its Workbook.

As a starting point Carrera and Ebeling Engineering suggested the following spring rates: 345#/" right front, 310 #/" left front, 165 #/" left rear and 150#/" right rear. This set-up on Mc Creary tires with the 131 compound produced a mildly understeering car on a flat oval. Mini Stock springs are 12 inches tall in the front and 14 inches in the rear. Carrera offers springs in a close range (10 to 20 pounds), which is an important factor considering the responsiveness to tuning changes of a lightweight car.

Elements which affect spring rate choice include: 1) total car weight, 2) aerodynamics, 3) form of racing, (asphalt or dirt) oval, road racing and track banking, 4) sway bar(s) rate(s), 5) motion ratios at the wheels, 6) roll center heights, and 6) front-to-rear and side-to-side weight distribution. Other factors such as power output and tire rule restrictions will also have a bearing on spring choice. Again, the chassis builder and shock manufacturer should be consulted for their recommendations on this subject.

Wheel Rates

Spring and roll bar stiffness mean nothing until those rates are computed into the leverage ratios supplied from the inner mounting point of the lower control arms to the center of the ball joint in the front, and the center of the rear end housing to the center of the rear tire contact patch at the rear. To calculate the wheel rates at the front, measure the distance from the inner lower control arm mounting point to the coil over (or spring) mounting point on the arm. Then measure the distance from the inner mounting to the center of the outer ball joint. Our car showed the following measurements: Lower control arm pivot point to coil over location 10.625 inches. Lower control arm pivot to center ball joint 13.50 inches. Divide 10.625 inches by 13.50 which equals .797. That figure squared is .62. So .62 is our front effective motion ratio. A spring of 345 pounds per inch, once installed in the Ebeling chassis, becomes: 345 times .62 equals 214 effective pounds per inch wheel rate.

The rate now has to be corrected for the spring angle. To do so, multiply the angle of the spring from vertical by the cosine of the angle of the spring. For example, 18 degrees from vertical equals .951 correction factor, so .62 times .951 equals .59.

In the rear, the wheel rates are computed by dividing the spring track by the wheel track, and multiplying that by the spring rate and the spring angle correction factor.

Once both the front total wheel rates are calculated (add together both front effective wheel rates and the anti-roll bar rates) and the rear effective wheel rates are

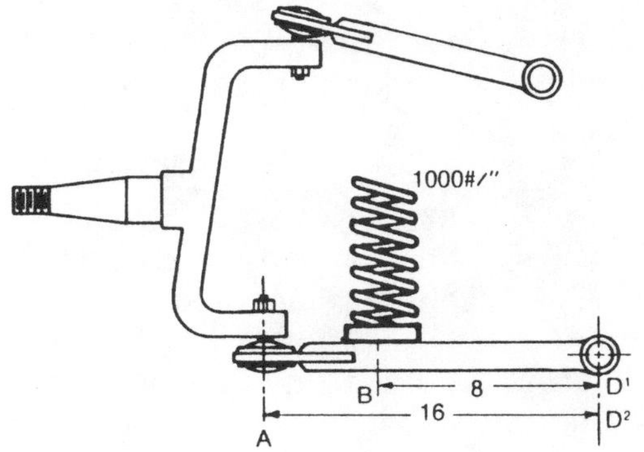

Wheel Rate

computed we can calculate the roll couple distribution.

Keep in mind that no matter how stiff the springs or anti roll bars are, the same amount of weight will still transfer from the inside to the outside tire contact patch. However, the relative rates of the front and rear springs will determine how much of that weight will be transferred at the front and how much will transfer at the rear outside tires.

Roll Couple Distribution

The roll couple distribution is the percentage of the effective wheel rate the car is carrying at each end during cornering. Roll couple distribution should be selected relative to the front-to-rear weight distribution and roll center locations along with all the other considerations that enter into the choice of spring rates.

A front-heavy car should carry a lower front roll couple than a rear-end-heavy car because a front-heavy car will tend to push due to the extra load placed on the front tires, which in turn increases their slip angle. A relatively stiffer rear spring rate will compensate for the heavier front weight by inducing some slip angle into the rear tires.

Let's say that the front of a car carries a total of 272 pounds of effective wheel rate and the rear 159 pounds, for a combined total of 431 pounds. We can now find the front roll couple by dividing the total front wheel rate by the total vehicle wheel rate. Thus we have 272 divided by 431 equals 63% front roll couple. Rear roll couple will be the difference between 100% and 63% or 37%.

Dynamic Weight Transfer

To compute weight transfer, use the formula —

$$WT = \frac{CGHxWxG}{TW}$$

where WT is total weight transfer, CGH is center of gravity height, W is total car weight, G is the gravitational force, (use 1.1 for all practical purposes) and TW is the average of the front and rear tracks. Using the values previously shown for our project car,

$$WT = \frac{11x2340x1.1}{61.875} = \frac{28314}{61.875} = 458 \text{ pounds.}$$

We now know that our car transfers 458 pounds during maximum (1.1G's) cornering, and we also know that the front end will transfer 63% of that weight, from our roll couple calculations. If the static weight distribution in the front is 732 left and 415 right and 63% of 458 pounds, 289 pounds, will move from the left front to the right front, during cornering we now end up with 443 pounds on the left front and 597 on the right front.

In the rear we follow the same basic format— 37% of 458 is 169 pounds of weight transfer. The left rear weighed 707 pounds, the right rear 488. With 707 pounds minus 169 of weight transfer, we end up with 538 pounds on the left rear and 657 on the right rear during maximum cornering.

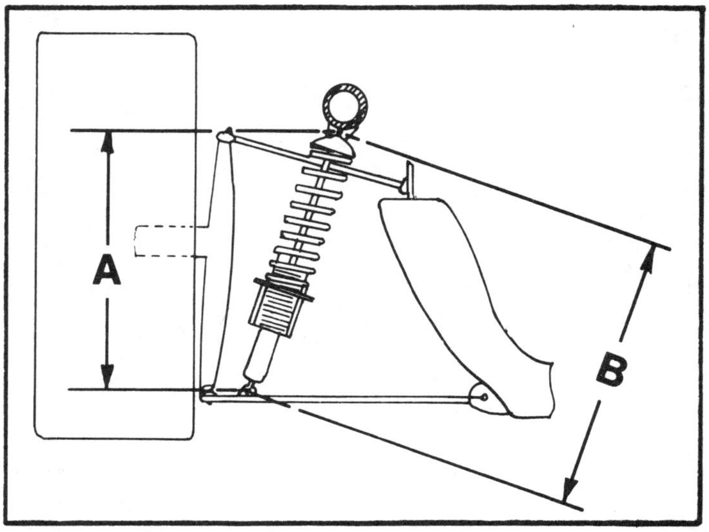

This is how the angle correction factor is determined for coilovers. "A" divided by "B" is the cosine of the mounting angle. Multiply the cosine by the motion ratio squared, then the resulting number by the spring rate, and you have the wheel rate. Be extremely accurate in making the "A" and "B" measurements, as a very small difference will introduce a very large error.

Static		Dynamic	
732	415	443	704
707	488	538	657

Notice that the total front weight on the static and dynamic figures is always 1147 pounds and the rear weight on the static and dynamic figures is also an equal 1195 pounds. We can also see from these figures that the weight difference between the front wheels is noticeably greater than the weight difference between the rear wheels. This condition could cause the car to understeer.

Weight Distribution

Correct weight distribution is an important element of a properly designed and properly handling car. The key is to coordinate the weight distribution with the roll couple so that each tire, during maximum cornering, is sharing an equal amount of cornering loads. The best example of this theory, is the Porsche racing car. Porsche is by far the most successful racing car ever, but the Porsche carries an undue amount of weight over its rear wheels. Porsches use a very high front roll couple to balance out the high rear end weight.

Since weight distribution and roll couple go hand-in-hand, it is possible to set-up the chassis by adjusting the weight distribution or resetting the relative roll couple.

This can be done by calculating the dynamic weight transfer with various roll couples and weight distributions. Cross weight can also be used to set-up the dynamic weight distribution.

High Left Side Weight

We saw earlier in some of our calculations that as the car corners, the inside weight diminishes and the outside increases proportionally. The ideal situation for a race car while cornering is to have all four tires equally loaded. If we start with a 50/50 side to side weight distribution, it is obvious that some of the weight on the left will end up on the right and the left will be under loaded and the right will be over loaded.

By starting with more weight on the left than on the right, if we are to turn left only, we will still transfer the same amount of weight to the right, but more of that weight will remain on the left and less will be on the right during maximum cornering. A well-designed short track car should be able to accomplish a 50/50 dynamic weight distribution (lateral) with proper suspension geometry, wheel rates and roll bars with a maximum of a 60/40 left-to-right starting point.

The higher the left side weight bias, the lower the front roll couple percentage required and the lower the overall spring rates required. This is due to the fact that less body roll takes place during cornering with a heavier inside load.

The inside weight to roll couple relationship works on a one to one ratio— one percent inside weight increase equates to a one percent front roll couple decrease. In other words, the more inside weight your car carries, the softer the front springs can be.

WARNING: Too much of a good thing can be bad. Too much inside weight can create a very sluggish right turn response and poor handling over bumpy surfaces. More importantly, over 60 percent inside weight can cause the car to overturn if a right turn is made or if the car is hit.

Front-to-Rear Weight Distribution

We said earlier that horsepower made a difference in the way a Mini Stocker was set-up compared to a more powerful V8 car. One of the areas where this is a factor is in weight distribution front-to-rear. It is generally accepted on a big car that the more weight on the rear the better, to improve traction under power. A four cylinder puts out a bit less than half the power and torque of a V8, so it stands to reason that the rear weight consideration is less important.

A bit more front weight bias can be used in a Mini Stocker, which allows a lower front roll couple to be used. The greater the rear weight distribution, the higher the front roll couple required. The generally accepted norm is 49 to 51 percent front weight distribution for a coil-over Mini Stocker.

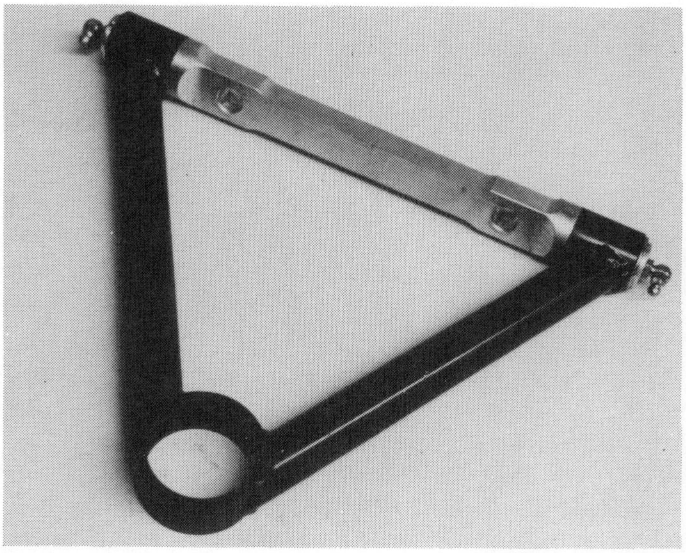

The Ebeling Mini Stock upper control arm. These arms are light due to their design where the two arms meet at the ball joint. The ball joint socket is welded directly to the arms.

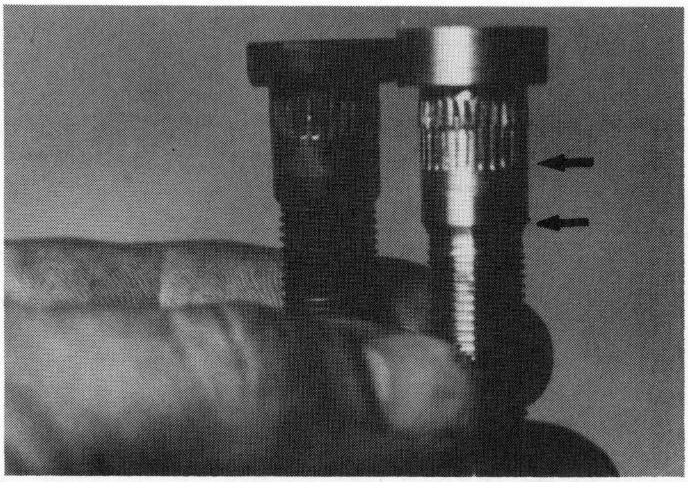

Wheel studs come in many different sizes. Make sure that even if the studs look like the stock ones, you check that the splines are not longer than the splines on the stock studs (top arrow) or the splines will crack the hub. Also check that the shank of the stud does not protrude beyond the face of the hub. If it does, the shank will crack the brake rotor hat.

This stock Pinto hub was turned down to accept the JFZ brake rotor hats. Lower arrow points to the crack caused by the splines on the stud. The shank on this stud is too long and it will interfere with the hole in the brake hat. Racing Pinto wheel studs can be obtained from ARP.

Chapter 7

The Rear Suspension

On any type of track, the key to performance is getting good bite and acceleration from both rear tires. This involves the interaction of weight transfer, induced downforce loading of the tires, and proper compliance at the tire contact patch.

The relationship between the track surface and the tire contact patch can be very fragile, especially when the tires are subjected to both lateral acceleration and forward thrust loading. This relationship can easily be destroyed by a harsh torque reaction or a violent suspension movement. The spring rates and suspension attachment linkages must be carefully designed to cushion the harsh and violent reactions, and preserve the fragile contact patch/-track relationship.

The Three-Point Rear Suspension

Our project car is equipped with a three point rear suspension linkage. The three point system is the traditional method of attaching the rear end to the chassis and it is the most widely used by stock cars and road racing sedans. The three point linkage includes two lower links 23 inches long, attached at each outer end of the rear end housing from a bracket on the housing straight forward to a bracket on the frame.

The third link is attached at the center on the top of the rear end housing, running straight forward to a bracket on the frame. The forward mounting height of this upper link can be adjusted to set the amount of anti-squat in the rear suspension. The three-point rear linkage is very easy to work with and it is highly adjustable so that it can be adapted to almost any type of racing condition.

The angles of a three-point link system, projected forward to a common point, form an instant center, and thereby actually behave like a torque arm. Because of this, be very careful in the layout of the three-point rear linkages so that this instant center does not have much angular movement during suspension travel. If the suspension

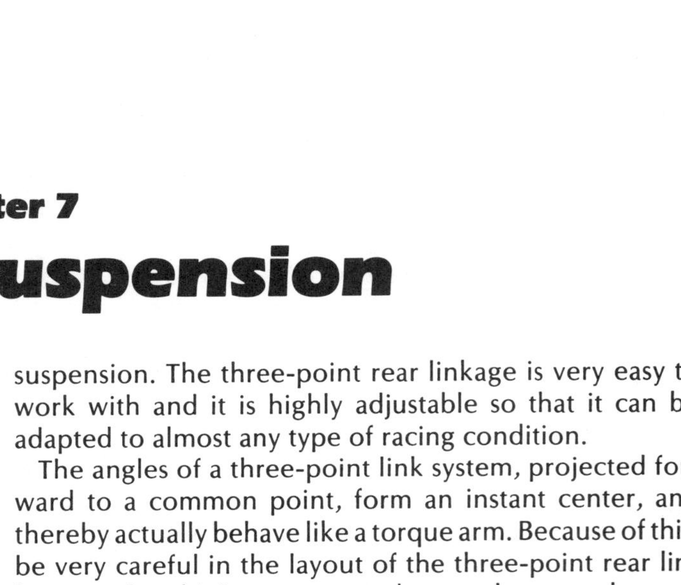

The top bar in the three bar system prevents the rear end from rotating around the wheels when the power or the brakes are applied. When the car accelerates the tires are applying torque in one direction and the rear end is trying to counter that torque in the opposite direction (to every action there is an equal reaction). Under power the front of the rear end is trying to lift and the back is pushing down, which causes a pulling action on the third bar against the chassis. The angle of the top bar in relation to the lower bars determines the anti squat.

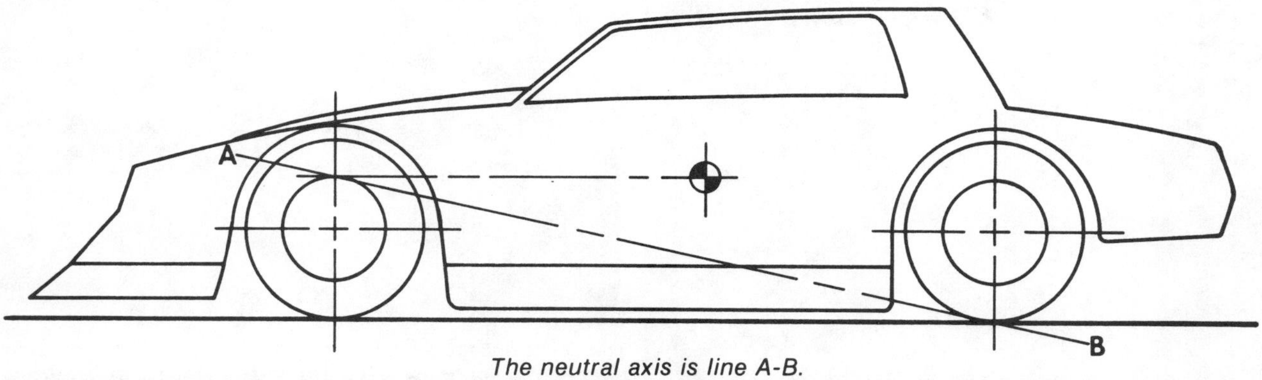

The neutral axis is line A-B.

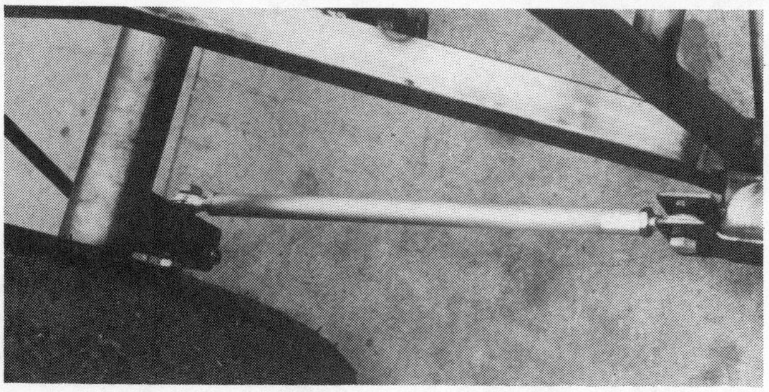

The lower bars should be mounted parallel to the ground for road racing. For oval track racing the car will probably be set-up with some lean to the left which will cause the right link to aim up toward the front of the car and the left link to aim down towards the front of the car. The opposite angles in the bars will help minimize roll steer if any is present in the rear suspension. Note the flats we machined in the bar for holding it with a wrench when tightening the rod ends. The flats were placed at the left hand threads so we would know which way to turn the bar when setting the rear end.

travel causes instant center movement, the rear wheels and suspension arms will be changing angles quite rapidly, changing the handling mode of the chassis.

If you are designing your own chassis, be sure to draw out the chassis and suspension linkage in quarter-scale before fabricating it. This will save you a lot of problems down the road.

The Neutral Axis

The neutral axis is an important point to keep in mind when establishing the rear suspension layout. The neutral axis is a line running from the center of the rear tire contact patch forward up to a point which is the intersection of a line drawn vertically through the center of the front wheel with a line drawn forward horizontally from the center of the center of gravity height (shown as line A-B in the drawing).

This neutral axis line is a plane of the car which is important to the rear suspension geometry in controlling acceleration pitch reactions of the vehicle. If the instant center of the rear suspension geometry falls on the neutral axis, there will be no dipping or raising of the car during acceleration. The reason is that the resulting action from suspension movement has no moment arm to apply a torque about the instant center.

If the instant center falls below the neutral line, during acceleration the rear suspension springs will be compressed by the leverage of the suspension arms and the rear of the car will squat, with the result that no or little downforce loading will be transmitted to the tire contact patches. This is a situation that should be definitely avoided.

If the instant center falls above the neutral axis, the rear of the car will move upward during acceleration. The best way to keep the car level under acceleration is to keep the instant center on the neutral axis or slightly above it. If the torque arm or the instant center of the rear suspension arms is attached to the chassis exactly at the fore/aft center of gravity and the neutral axis, the entire car will uniformly rise under acceleration, and pitching will not occur.

The Panhard Bar

The full length Panhard bar is the simplest and most widely used of the lateral control linkage systems. It is attached to the chassis at one side of the car and at the axle housing at the opposite side of the car. The Panhard bar is simply a long tube with spherical rod end bearings at each end which braces the rear axle from one end to the frame at the other end, resulting in no side shift of the axle.

A drawback of this system as compared to the Watt's linkage is that it moves in an arc rather than straight up and down. The shorter the bar, the sharper the arc it scribes.

Using a full length bar which goes all the way from one side of the chassis to the other, the amount of arc it moves in is very minimal if the wheel moves a normal distance into bump (up to 2 1/2 inches). With more bump traveling significant bump steer. This geometry however, the increased arc of the bar will begin producing significant bump steer. This geometry problem is accentuated when the bar is shortened.

The long Panhard bar is the way to go for economy. The only problem on a Mini Stock is that the bar has to be bent to clear the rear end quick change cover and it has to be reinforced to not bend under cornering loads. The advantage to the Panhard bar is that it will flex if the rear wheels get hit. The Panhard bar has the added advantage of allowing the roll center to be changed by relocating its mounting point height.

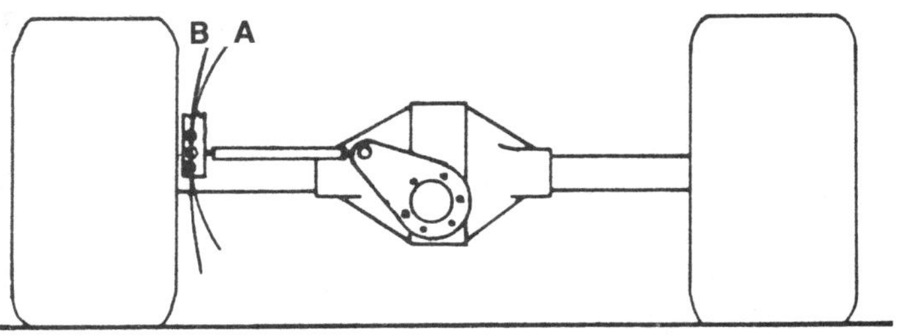

"A" is the arc scribed by the short Panhard bar illustrated. Compare the sharp arc of it to the smoother arc of a long Panhard bar, illustrated by "B."

The Panhard bar must be mounted parallel to the ground to prevent its arc from moving against the natural bump arc of the outside wheel, creating bump steer by forcing the whole axle toward the outside of the turn.

It is extremely important that the bar be straddle mounted between brackets which are stiff enough for the application. Most racers do not realize the extreme torque which is placed on the Panhard bar brackets. The brackets should be triangulated and welded correctly.

The Watt's Linkage

The ideal car has zero roll steer and bump steer which prevents it from feeling "twitchy" in turns and over bumps. A Watt's linkage has a bellcrank which is attached to the center of the rear end housing on a pivot. It has two arms from the bellcrank, one extending in each direction, ultimately anchoring in a bracket on each side of the chassis. Each end of these arms is free to pivot in spherical rod end bearings.

The system keeps the chassis centered in its proper position over the axle during all movement of bump and rebound, without any bump steer resulting from a linkage

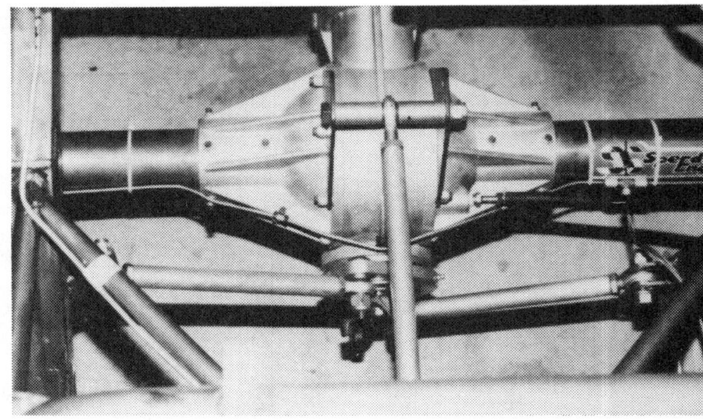

The Watts linkage system is a better locating device because of its rigidity. The only drawback to the Watts linkage is it is so strong that if the rear wheels get hit, the bellcrank can break the rear end center section. The Watts linkage system has a set roll center at the center of the rear end pinion.

arc moving the outside wheel through a forced arc.

Watt's linkages are available from Speedway Engineering. We chose to use a Watt's linkage locating device for our project car.

Anti-Squat

Anti-squat is a very important element in both a paved and dirt track chassis to help gain better rear tire traction. As a vehicle accelerates, weight is pitched rearward in reaction to the vehicle's forward movement. This rearward pitch of the vehicle weight can be reacted in two ways at the rear suspension. The first case is with an anti-squat geometry in the rear linkages. Here the weight is utilized simply by the springs using up compression travel. No additional weight is transferred downward onto the tires.

The second case is a car with anti-squat. Anti-squat utilizes the linkage arrangement of the rear suspension to brace the rear end housing against the body moving downward on it. The linkage actually acts as a lever to pry the rear end housing away from the body as the body pitches rearward, and forces the body upward. The leverage action causes additional downforce to be placed on the rear tires.

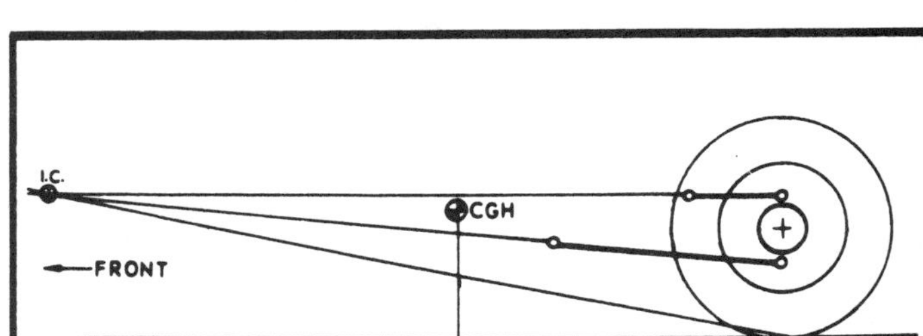

Anti-Squat

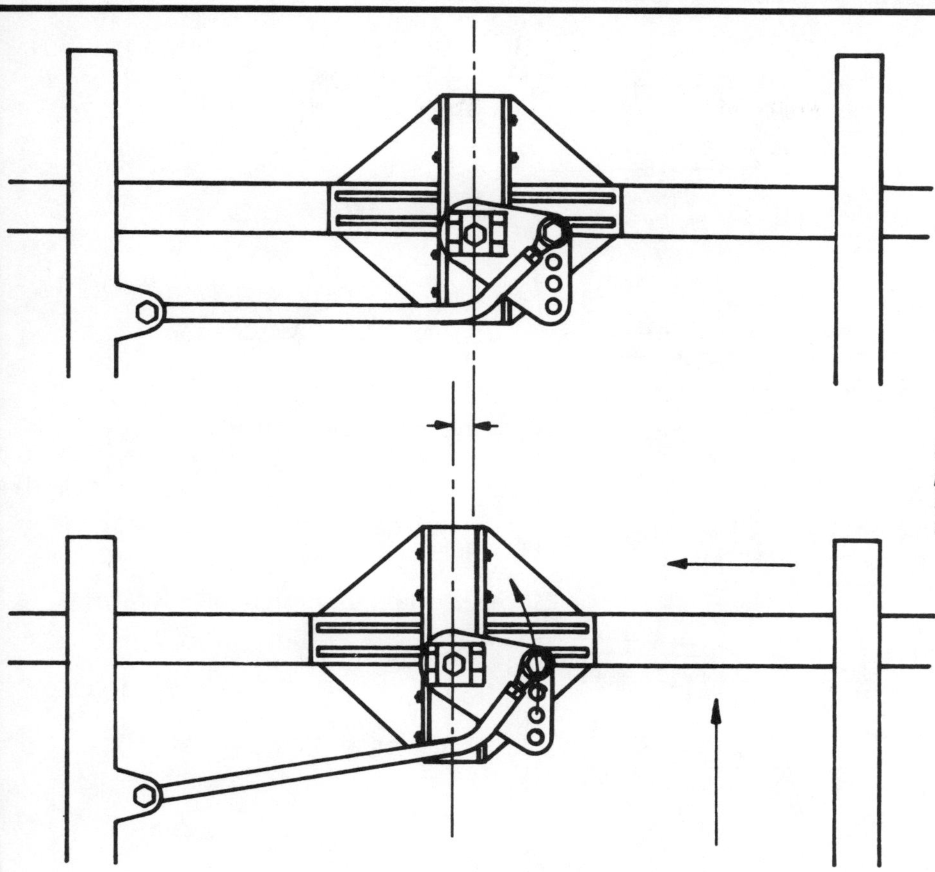

This demonstrates how a short Panhard bar creates lateral displacement and bump steer during body roll or one-wheel bump travel.

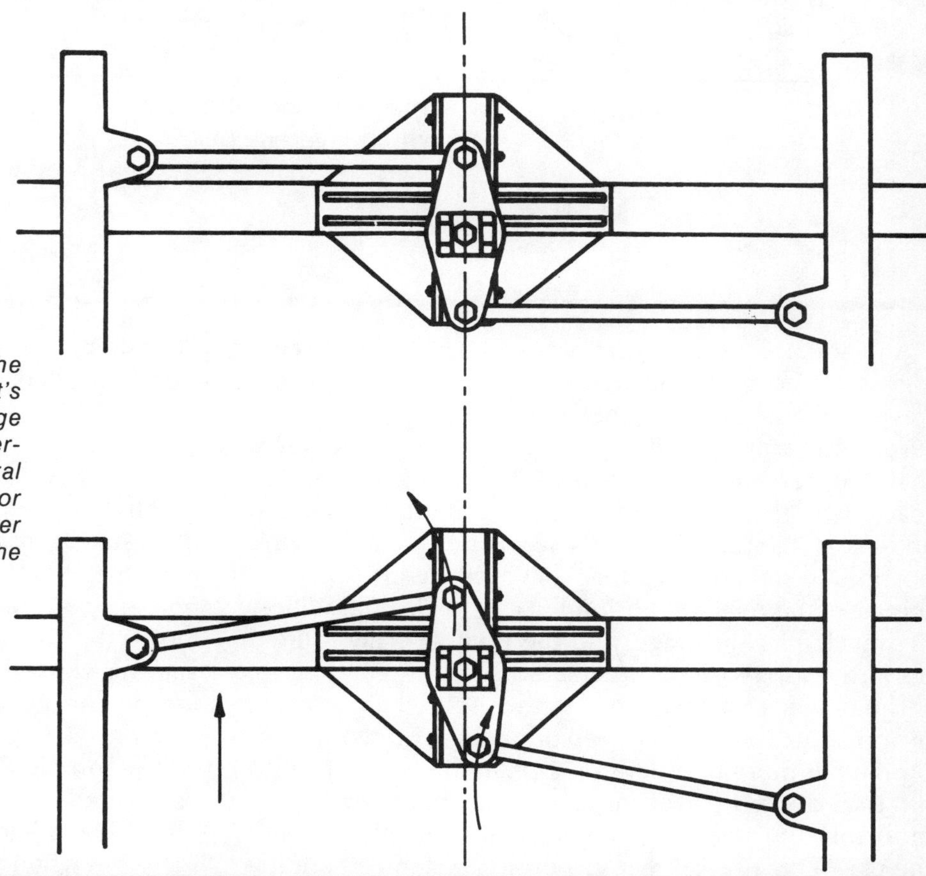

A demonstration of how the Speedway Engineering Watt's linkage for the quick change housing keeps the housing perfectly centered with no lateral displacement through body roll or one-wheel bump travel. Another feature of this system is that the roll center is located quite low.

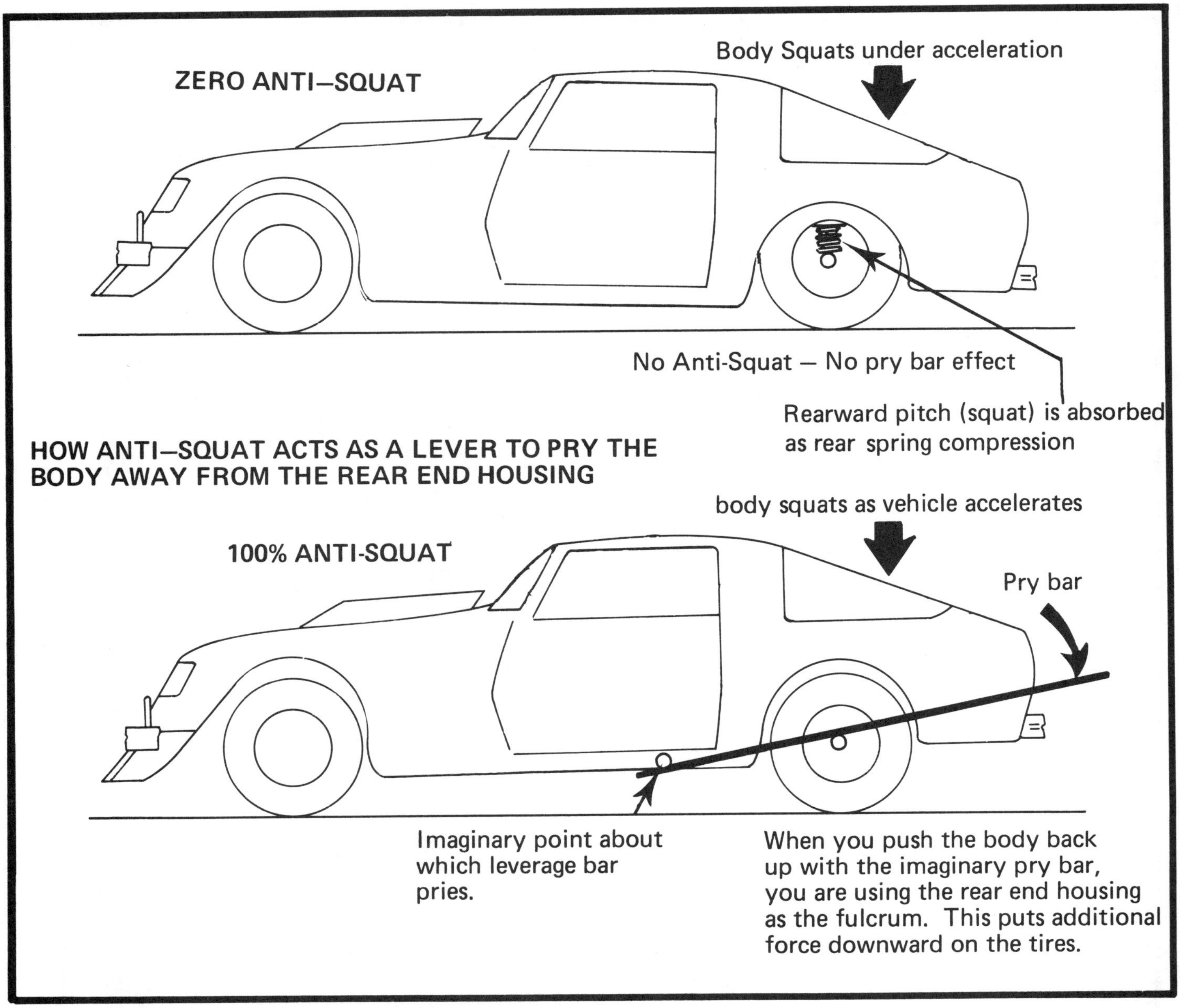

Anti-squat is expressed as a percentage. Fifty percent anti-squat means that 50 percent of the rearward pitching travel is reacted by the mechanical leverage of the rear suspension arms. One hundred percent anti-squat means that 100 percent of the rearward pitching moment is reacted by the mechanical leverage. With 100 percent anti-squat, the rear of the vehicle will not squat at all. It is possible to obtain more than 100 percent anti-squat. In this case, the rear suspension linkage begins to multiply the mechanical leverage, and the body actually is lifted upward during forward acceleration.

The drawback associated with the use of high amounts of anti-squat is rear wheel hop under heavy braking. This problem is more prevalent on asphalt than on dirt, as on dirt the sideways drift at corner entry tends to slow the car more than the brakes do. This drawback is eliminated with the use of a sixth coil rear suspension system, where the undesirable forces in the rear linkages under braking are well controlled.

Roll Steer

Roll steer is an element induced into the dynamic handling state by the longitudinal movement (along the car's wheelbase) of the solid rear axle as the car moves through body roll. It happens as an axle attaching trailing arm moves through an arc, changing its effective length. Linkages can create roll oversteer or roll understeer, depending on the geometry of the linkage. Neither of these elements are desirable. Measurements should be taken on the completed chassis to determine its existence in the completed suspension.

Roll oversteer will occur if the right side wheelbase increases when the car is turning left, and if the left side

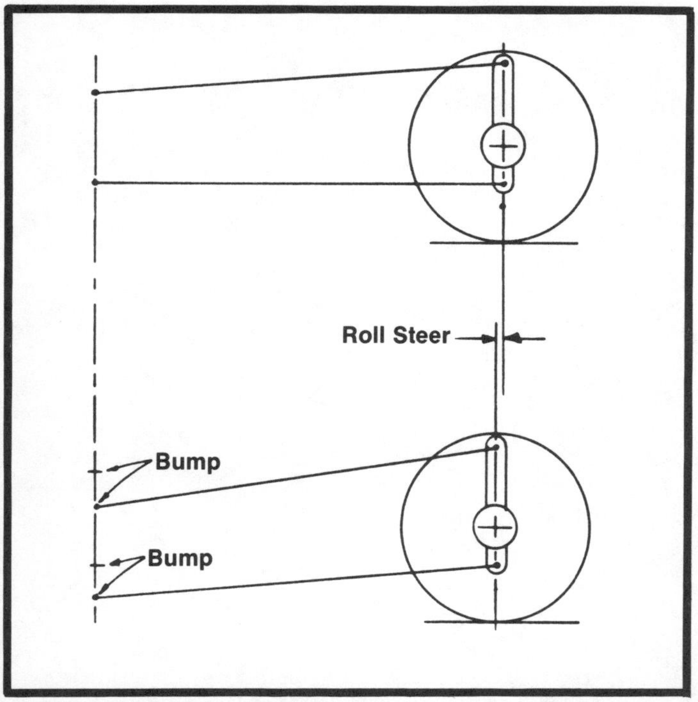

Rear roll steer with 3-point linkage.

wheelbase increases when turning right. What happens is that the rear wheels are now trying to power steer the rear end. If the rear end is facing the inside of the turn (longer inside wheelbase) when power is applied, the rear end will try to drive towards the inside of the turn, causing understeer. If the rear end is aiming toward the outside of the turn (longer outside wheelbase), the rear end will try to pull the rear of the car toward the outside of the turn, causing oversteer.

Using Roll Steer

Roll understeer should be avoided. It is difficult to balance the chassis with it present. A chassis that, unknown to the racer, has roll understeer present usually is adjusted by increasing oversteer, probably by adding large amounts of stagger. But this induces a new problem in the chassis to cover up an existing one.

A small degree of oversteer can be helpful, especially with a car which has a solid locked rear end. It works in a positive way the same as a small amount of tire stagger does. It can be most helpful on a dry slick dirt track.

Rear Coil-Over Mounting

The bottom mounting point of a coil-over unit on the rear end housing should be located eight inches below the centerline of the axle tube. The coil-over unit should be mounted behind the rear end housing.

The bottom coil-over attachment point should be located five inches from the inside of the face of the rear rotor on each side. This has become the industry standard so that rear spring rate numbers will relate from one car to another with some consistency.

The Rear Trailing Arms

The rear trailing arms and Watt's links are made of 1-inch OD x .156-inch wall 2023 T-3 aluminum. Aluminum has nearly the same yield point and tensile strength as steel, but it weighs about one quarter of steel.

We used Aurora 5/8 inch rod ends on the three arms and the Watt's link. The fasteners are F911 series from ARP. The rod ends are from Baker Precision.

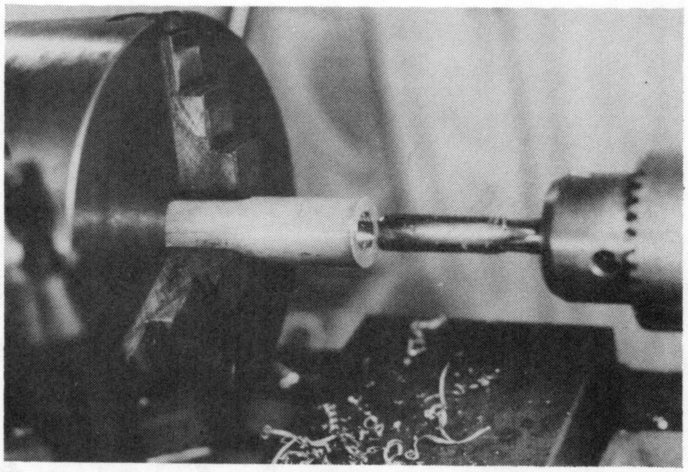

The threads in the bars should be cut on a lathe. One end is right hand and the other is left hand to allow adjustment by simply turning the bar, without removing the rod ends. A mark on the wrench flats should be placed on the left end of the bars so anyone who works on the car will know which way to turn the bars when they need to be adjusted.

In the rear, the coil-over adjustment sleeves will have to be set on the upper notch on the shock to allow full adjustment. Use some type of limiting straps or chains to limit rear end travel.

Chapter 8

Chassis Adjustment and Troubleshooting

Basic Set-up

The very first thing to do before setting up the chassis to get a note pad and a pencil. The figures taken offthe car must be noted to be referred to at a later date when the set-up will need to be changed to alter the handling. Another important reason to keep such data is checking the chassis ride heights and camber at the track to see how the surface you are now working on varies from your garage floor.

For example, let's say you had 1 degree negative camber in the right front and 1 degree positive in the left. You get to the track and the camber now shows up as only 1/2 degree negative on the right and 1/2 degree positive on the left. This means that the surface you are working on is tilted to the right 1/2 degree. That's OK— as long as you park the car on the same spot and in the same direction each time you come in the pits. Experienced racers actually mark their spot with nails so they can reuse their pit race after race with the same results.

A lot of calculating and plotting should go into the setting up of a race car. Steve Smith's Work Book for Advanced Race Car Development and Advanced Race Car Suspension Development were very helpful in setting up the Mustang.

Chassis Set-up Data

Here is the set-up we started with on our Ebeling chassis Mini stocker. Track is a 1/3-mile asphalt, flat to off camber turns.

Front track: 62 inches   Rear track: 60.125 inches
Wheelbase: 96 inches
Weight distribution: 49% front, 51% rear
60% left, 40% right
48% cross
Corner weights: LF—708, RF—438, Total front— 1146
LR—696, RR—498, Total rear— 1194
Inside— 1404   Total— 2340

These weights are with driver on board based on the rules at our home track.

Ride heights: Left front— 3½″   Right front— 4″
Left rear— 4″   Right rear— 4½″
Camber: Left— 1 degree positive
Right— 1 degree negative
Caster: Left— ½ degree negative
Right— 1 degree positive
Rear stagger: 1¼ inches
Front stagger: 1 inch
Rear end: Square
Tires: Mc Creary 131 compound, 23.5/10-13S
Pressures: Left front—13   Right front—18
Left rear— 13   Right rear— 19
Springs: Left front— 310   Right front—345
Left rear— 165   Right rear— 150
Shocks: Carrera 5176 (set on number 2— 50/50) on all four wheels.

The right rear trailing arm angles up toward the front slightly and the left rear arm angles down slightly to anti-squat.

Gearing is 4:11 with a 17 Speedway Engineering gear for an overall ratio of 6.85.

A few things would have made some difference in lap times: 1) Gearing to 200 to 400 RPM past the power peak (7500 RPM) would have been worth another couple tenths; 2) aerodynamics did not seem important at the time, so I didn't build a front air dam and a rear spoiler, but it was later realized that the car was coming off the corners at nearly 65 miles per hour and the straights brought speeds of up to 75 MPH. The average speed of 70 MPH definitely warranted a lot of work on the aerodynamics.

One of the peculiarities of a Mini Stock is the lower power output, so that air turbulence was hurting the Mini Mustang a lot more that a V8-powered car would have been. The front of the hood was about 12 inches away from the ground allowing a large amount of air to enter under the car, causing both lift and drag. An air dam and radiator shroud are in the works from Limni Racing. Aerodynamics was probably part of a problem that was experienced with the engine temperatures being elevated. Air under the car was probably causing a high pressure area behind the radiator, preventing cool air from flowing through the radiator.

A front air dam such as the one on our project car reduces air drag and front end lift. Cooling is also improved because the air is forced to go over the hood into the radiator opening and by being kept out of the underside of the car it creates a low pressure area that sucks the air through the radiator. The material used to make these air dams can be purchased from Limni Racing.

This set-up placed the car in the trophy dash in just it's second outing with a rookie driver on board. The gearing was off by about 500 RPM. Reading tires after the main indicated too much negative camber in the right front and too much positive in the left front, probably causing the push that was experienced in the turns.

Chassis Set-up Procedure

With the car set up properly in your garage, you should have to make only small (if any) changes to the chassis when you get to the track. Before doing anything, study this set-up flow chart:

1) Disconnect anti-roll bar
2) Set the tire pressures, stagger and ride height
3) Set corners to desired weights
4) Square rear end
5) Check wedge
6) Set caster
7) Set camber
8) Set toe
9) Neutralize anti-roll bar

The reason behind this sequence is that setting one thing will often upset another adjustment. For example, the toe will change if it is set before any of the adjustments on the flow chart. In fact, if the chart is not followed, the only setting that will remain correct will be the last one done.

1) Disconnecting the anti-roll bar is essential before doing any chassis adjustments. If the bar is not neutral from side to side, it will put in some pre-load on the chassis, negating the work you are doing. Even if you were to start the set-up session with a perfectly neutral bar, any change made to the chassis that would upset the ride heights would put some pre-load in the bar, and even caster and camber will have an effect on bar pre-load.

The Flexi Flyer scales (above) use a hydraulic piston and a pressure gauge. The Speedway Engineering chassis checker (below) is a lever arm that is placed in the rim lip and is supported by a fulcrum. A torque wrench is used to apply just enough force to lift the wheel off the ground. The reading on the torque wrench is multiplied by 10 to get the weight at that corner.

2) Setting the tire pressures and the stagger will ensure that the tire diameters are at the same dimensions they will be on the track. A different size tire will change corner weights and ride heights.

Ride height is set by trial and error. But one rule of thumb should be kept in mind: changing one corner will also change the diagonally opposed corner by the same amount. For example if you add 25 pounds on the right front by screwing the coil-over adjuster in, the left rear should also increase by 25 pounds. "Should increase"— this just never seems to work out, but there will be some change on all four corners when one corner is changed. If one corner height is changed, most likely the other three will also be affected. The easiest way I found was to set the front and rear separately and then play with each corner individually.

After making changes, the car should be rolled back and forth a few times and the suspension should be settled by bouncing on the bumpers to restore the shocks to their neutral setting and to allow the front wheels to go back to their correct position. A trick that saves a lot of time is to jack the car by the lower control arms before the corner is set back on the ground— this will take up most of the suspension travel and maintain the camber in the wheel.

3) The corner weights can only be set after the ride heights have been adjusted. To set the corner weights you will need a set of scales or a weight checker such as the one sold by Speedway Engineering. The best set of wheel scales I found are made by Fast Fab in Klamath Falls, Ore. These scales use four lever arms and bathroom scales. The Fast Fab scales are very accurate, but they are also very time consuming. The car must be taken off the platforms each time a change is made, and the levers must be re-set squarely under each tire or the readings will be off. I used the Speedway Engineering weight checker at the track and for setting the chassis at home and the Fast Fab scales for fine tuning.

You will encounter the same problem when setting corner weights as you did when setting the ride height — the weights in the chassis just never seem to do what they are expected to do. It just takes a lot of patience and going back and forth, checking all four corner weights and ride heights each time a change is made.

To maintain the pre-set ride heights change all four corners the same amount of turns on the coil-overs. For example, if you need to increase the weight in the left front and right rear, go up two turns on the left front and right rear and go down two turns on the right front and left rear. Reset the ride heights if they change before working on the corner weights again.

4) The rear end should only be squared after the ride height and corner weights have been set because the ride height offset from side to side may have an effect on the rear end alignment.

The rear end is most easily squared by stringing the car. Stringing a car is an exercise in basic geometry and the rule to remember is this: if two lines are parallel to each other, a third line parallel to either line will also be parallel to the other, and so on. So if we run a string parallel to one of the frame rails and a second string on the other side of the car parallel to the first string, we will have two strings parallel to each other and parallel to the frame rails.

Now measure the distance between the string and the rear tires at the front and rear of each tire at the bulge. The distance between the right tire and the right string should be the same at the front and the rear of the tire (unless you want some rear steer) and the distance between the string and the tire on the same locations, at the bulge, should also be equal on the left side. The actual distance from the string to its respective tire is immaterial as long as each string is parallel to its tire and the frame.

Do not be alarmed if you find that one wheel is toed out and one is toed in. Minor misalignment of the wheels, tires or even in the rear axle tubes will add up. Set the rear end so both wheels are toed in or out the same amount and in the same direction (both in or both out).

When turning the trailing arm bars to square the rear

end up, remember that even 1/4 or 1/8th turn will show up on the tape measure. 1/4 turn on one of the bars is enough to change the handling. Be sure to tighten the rod ends and double check the setting after everything is tight.

A small amount of rear toe steer may be beneficial on a dry dirt track. The rear end would have to be set up with the left side toed in and the right side toed out. This will cause the car to loosen up some at corner exit. 1/16- inch is about the maximum of rear end misalignment you would ever want to dial in.

Keep the strings in position— you will need them for the front end alignment.

5) If there is more than 50 percent left side weight bias in a car, the wedge cannot be checked in the conventional way (placing a jack in the center of the rear end housing with a pivot point on the jack such as a socket, and jacking up the rear end until the second tire begins to clear the ground). Our car has 60 percent left side weight bias. Whether the car has wedge in it or not, the right rear wheel is going to come off the ground first because the left side is heavier.

The best way to check wedge on a weight biased car is to jack the car up at the center of the weight distribution, not the dimensional centerline of the rear end. Let's say the rear track is 60 inches and 60% of the weight is on the left side of the car. Even though the dimensional centerline is

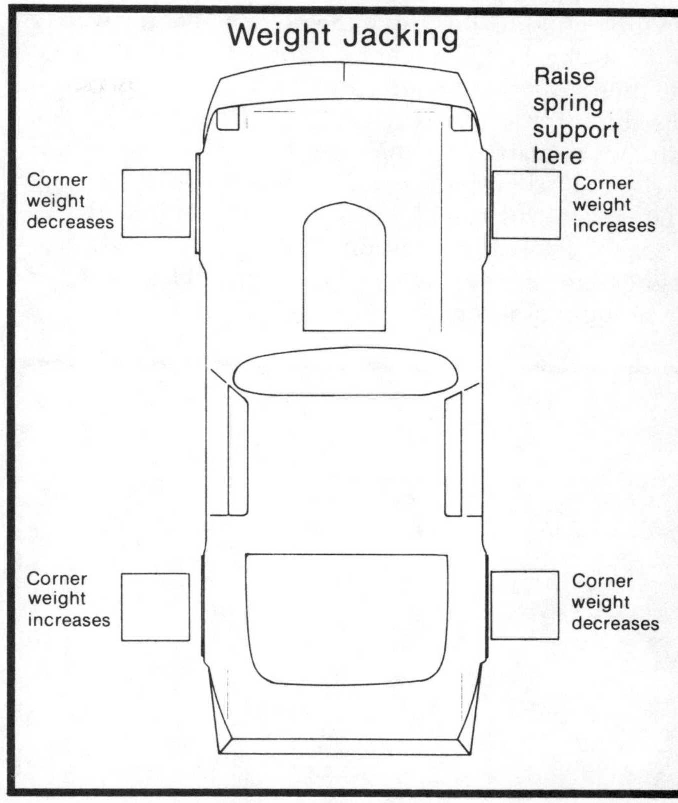

Weight Jacking Example

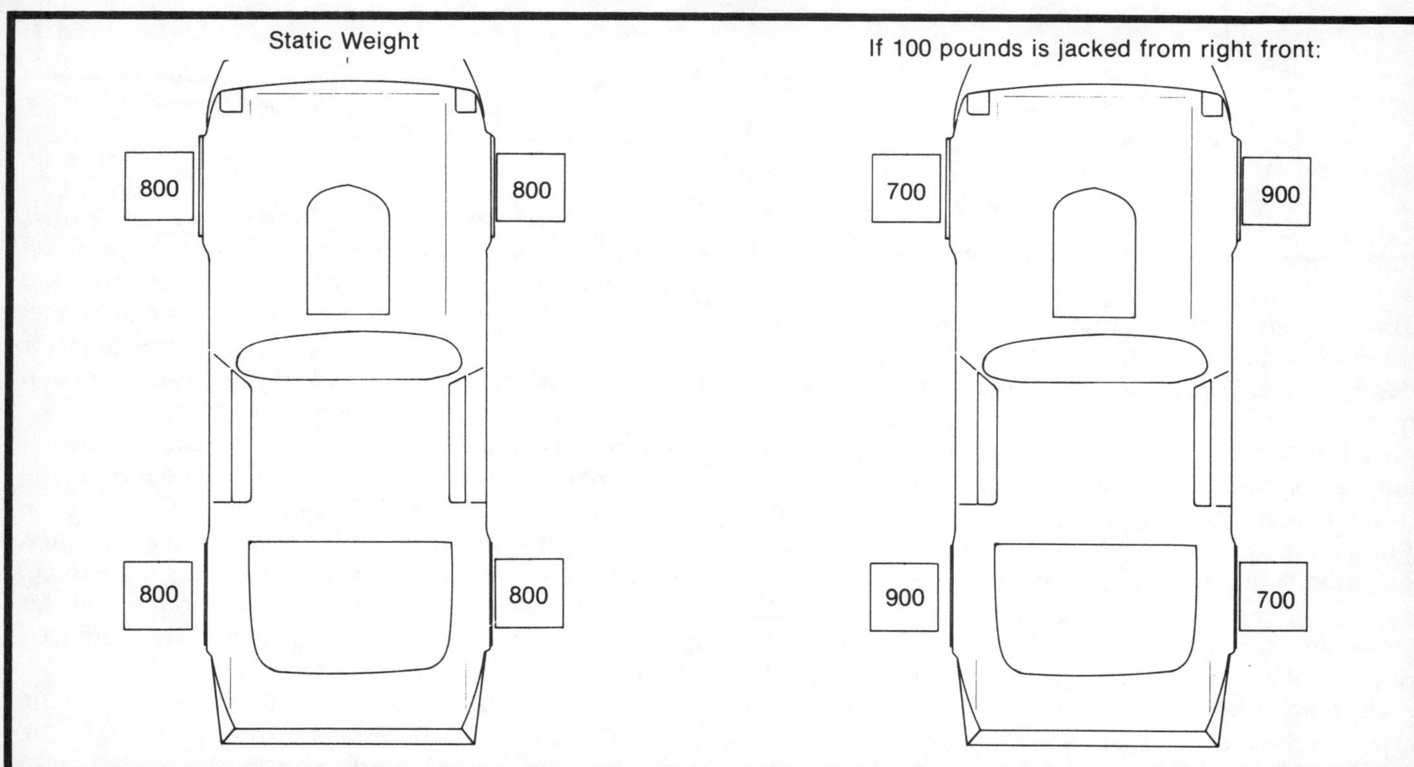

Notice that: 1) The total left side weight is still the same; 2) The total right side weight is still the same; 3) The total rear weight is still the same; 4) The total front weight is still the same. This assumes that the chassis is absolutely rigid.

30 inches, the center of the weight is 60% of the distance from the right side tire. This is the point where the wedge should be checked. It might be a good idea to weld a flat plate there to make the job easier.

Use this formula to calculate the location of the jacking point on a car carrying some weight bias; Track width x % bias right side dimension minus track left side dimension. So, $60 \times .60 = 36$ (the right side dimension), and $60 - 36 = 24$ inches, the left side dimension. The higher the right rear tire goes before the left rear clears the ground, the greater your amount of wedge is.

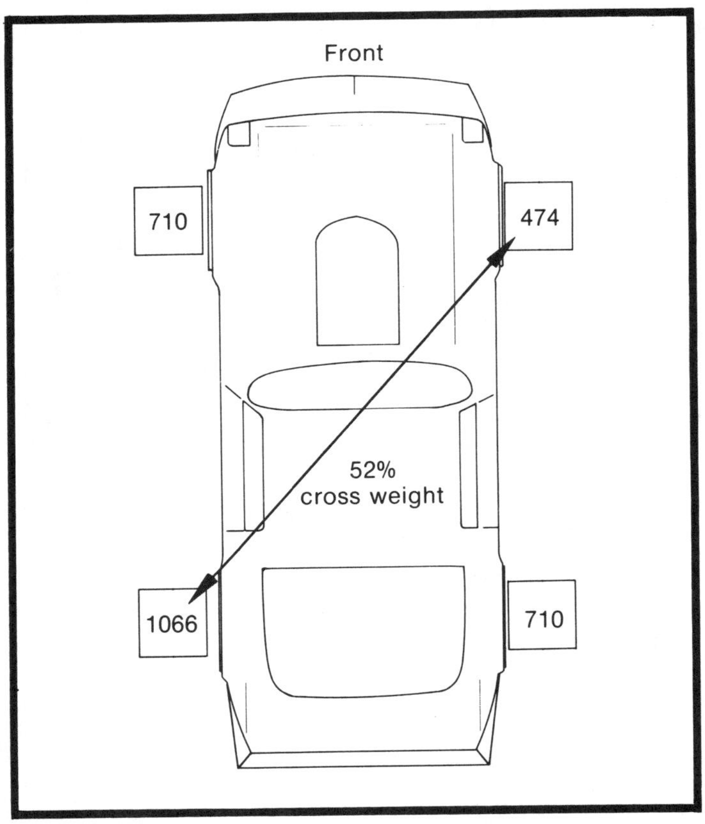

Wedge is related to cross weight percentage. The greater the cross weight percentage, the greater the wedge will be. When the wedge is changed, the cross weight will change by a proportionate amount. Jacking weight into the right front, left rear, or both, will increase wedge and cross weight percentage. Reducing the weight at those corners, or both, will have the opposite effect.

I find it much easier to work with cross weight percentage rather than wedge. Cross weight percentage is more meaningful than wedge measurement. The cross weight percentage is derived by dividing the sum of the right front weight and the left rear weight by the total vehicle weight. For example, in its second race, our Mustang had: 438 pounds on the right front, and 696 on the left rear for a total cross weight of 1134 pounds. 1134 divided by the total weight of 2340 pounds gives us 48%. Cross weight is what is set when adjusting the corner weights. The following will help in setting your car up with cross weight:

Situation:	Car needs:
Car pushes	Less cross weight
Car is loose	More cross weight
Spoiler added in rear	Less cross weight
Front air dam added	More cross weight
More front weight added	Less cross weight
More rear weight added	More cross weight

Remember that cross weight is always computed from the right front and the left rear.

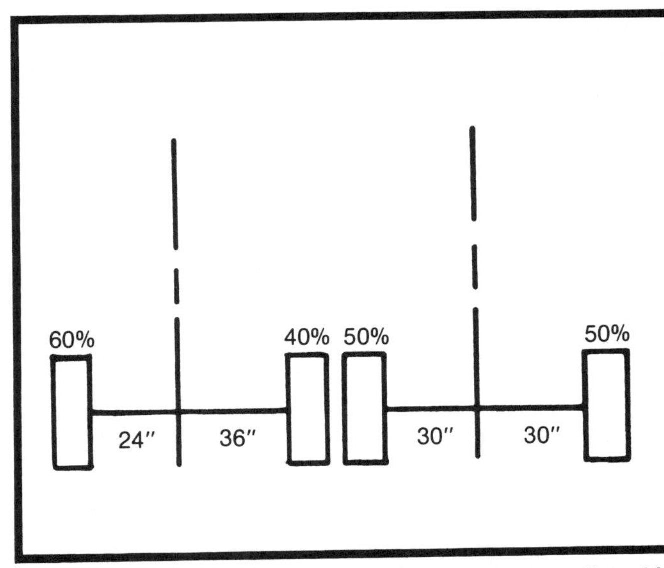

The center of weight distribution is no longer the same as the vehicle dimensional centerline when there is left side weight bias.

6,7) Setting caster should be done before the camber because camber is affected by caster. Caster is easily set with a gauge from Wayne Mitchell Engineering in San Diego, Ca. The gauge has a water level with a dial to level the liquid. When the liquid is leveled, you readthe dial for the camber or caster setting. Avoid the raised letters on the tires when using this gauge to get a true reading.

Before you start on the caster setting, place a piece of cardboard under the front wheels to facilitate turning the wheels from side to side. The wheels will have to be turned to one side to take the first reading for caster, then the wheel is turned to the other side for a second reading. Mitchell supplies a set of instructions to follow with his gauge, along with a simple calculation to compute true caster.

Stock cars use caster stagger. Caster stagger is nothing more than a difference in stagger from one side to the other. Our car uses 1/2 degree negative on the left and 1 degree positive on the right, for 1 1/2 degrees of caster stagger. The caster stagger helps the car turn left, but it

This camber/caster gauge is sold by Mitchell Engineering. It is easy to carry to the track and it is relatively inexpensive. Here the gauge is being used to check the camber during skid pad testing. Note how smooth the tire is, indicating much too hard a compound.

This tool is used to scribe a line on the tire tread surface to set front end alignment. Jack up the car and spin each front wheel and push the tool on the tire tread surfaces. Put the car back down and roll it back and forth to set the front end back to ride height. Use a long trammel bar or a tape measure to check the distance between the two tires.

also makes it hard to steer to the right if the driver needs to make a quick evasive maneuver on the track. Caster stagger is also used to reduce left front wheel loading while turning left, a situation which also unloads the left rear.

8) The toe should only be set with the steering rack in the centered position. If you make a tool that can be wedged in between the rack housing and the tie rod knuckle you will have a quick and easy method of centering the rack before checking the toe or any of the other settings. Keep in mind that just a slight amount of steering angle will change the toe. Not only is this confusing but it will also cause a wrong setting in the front wheel alignment.

To set the toe when the strings are still up from doing the rear end, just follow the same procedure used in the back. Set each wheel at 1/2 of the total toe desired. If you want 1/8-inch total toe out, set each wheel to toe out 1/16-inch. 1/16-inch will be the difference read off the tape measure between the front of the tire and the rear. The tire side bulges are the easiest place to use for measuring.

Another method which is not quite as accurate as the stringing is to measure the distance from the frame to the front and rear of each tire and again the difference in the two readings is the toe at each wheel. Scribing a thin line on the tire surface with a screwdriver by spinning the wheel will make an easy reference point to use.

The easiest method that can be used at the track is to use a "toe bar". The toe bar has a pointer at each end. One pointer is set against the side of a front tire while the distance at the other tire is measured and noted. By measuring both the front and the rear of the front tires and calculating the difference in measurements, the toe figure can be arrived at.

These solid bars in place of the coil-over units make wheeling the car around the shop easy. Cut the tubes to attain racing ride heights so all fabricating and welding can be done at a realistic level. These bars should be used when calculating the center of gravity height as shown in Steve Smith's Work Book for Advanced Race Car Suspension Development to prevent the suspension from compressing when the car is jacked up at one end.

9) Reinstall the anti-roll bar, but be sure no force is used to do so to prevent pre-load. Loosen the adjustment turnbuckle and screw the rod ends until the mounting bolt slips freely into its hole. If you ever want to check if any pre-load has been set in the sway bar, grab one of the arms, at the rod ends, and move it gently up and down. If there is no pre-load you will be able to feel the play in the rod ends just with finger pressure.

Setting Bump Steer

Bump steer is checked when the chassis is first assembled or after the car has been crashed. The bump steer is most easily checked using a gauge produced by Vic Irvan Enterprises. The major difficulty in checking bump steer is making a stand that is stiff and straight enough. The Irvan gauge is reliable, it provides accuracy and it is fairly inexpensive.

To set bump steer, first set the alignment within a reasonable limit, place the car on jack stands and disconnect the sway bar, springs and shocks. Remove wheels and bolt the hub plate provided by Irvan or your home made plate. Now place a jack under the lower A-frame and level the hub plate.

At this stage I found it easier to set the hub at ride height and check the toe change through two inches of bump and then go back down to check the change through an inch of rebound. This way I had a reading below and above ride height, with ride height being zero. Adjust the dial indicators on zero and set them level on the plate. Jack in one-half inch of travel at a time and note the dial indicator readings. For circle track racing, as much as .050-inch is acceptable, while for road racing as close to zero as possible is preferable.

With the initial set-up there was .040-inch of bump and, after shimming the tie rod end up about 1/8-inch using the Baker Precision insert and rod end, the bump came down to about .030-inch, but the readings increased from an inch travel. Moving the rack forward about an inch smoothed the curve of bump gain and reduced the bump to about .015- inch. The remaining bump was mostly at the travel extremes. For oval track racing there is no need to be so concerned about removing a small amount of bump steer, but we wanted to investigate the situation to report the findings in this book. Road racing bump steer is more critical because those cars generally see more bump and rebound travel.

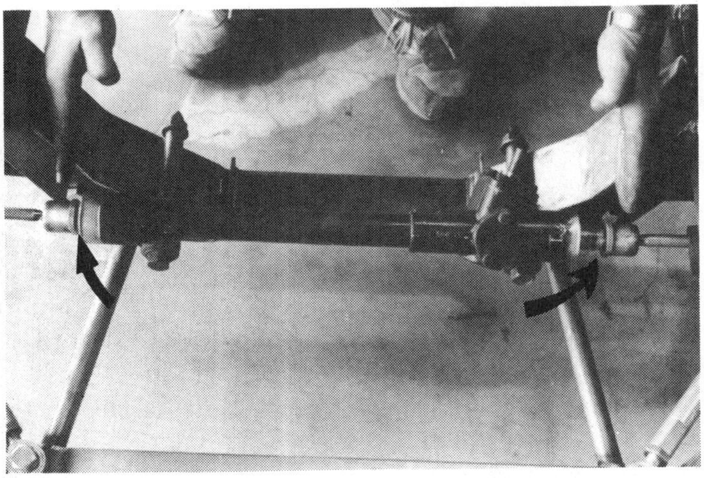

When aligning the front end, the steering knuckles should be in line with the lower control arm bolts. Note the different distances between the steering knuckles and the rack on both sides even though the rack is centered.

This Mini Stocker is leaning too much in the turn. This car would probably benefit from stiffer springs. From the looks of the front wheels, the car is pushing severely and it probably needs stiffer rear springs. Also the large rear spoiler probably compounds the problem.

This Ebeling Mustang is also understeering.

Adding Ballast

In its unbalasted form our Ebeling chassis car weighed about 1600 pounds. Our Mini stock association requires one pound per CC, with the driver on board. We allowed 200 pounds for the driver and fuel. Since our motor had to be cleaned up to 2340 CCs, 540 pounds had to be added. 540 pounds of lead is a lot, but we can use that situation to our advantage by placing the lead in selected areas to increase the inside weight and the rear weight bias. Check your rules for inside weight and total weight guidelines.

Chances are that most of the weight will have to be added to the left rear, near the battery. Think about installing the battery behind the left front wheel — the space at the rear can be used for ballast, which is much heavier than the battery for the same amount of space. The shorter cable to the starter will also help in reducing voltage drop.

First weigh the car and the lead you will need. Set as much lead as you can near the left rear and keep an eye on

This tray on the left side of the chassis made lead positioning easy. The tray is welded to the bottom of the frame to lower the center of gravity.

the scales as you keep adding. When you get to the point where you have achieved your desired weight distribution, start moving the weight around but try to keep it all in one area to reduce the polar moment of inertia.

For example, if you have to add some weight to the right, do not place the lead all the way over to the right side of the frame. Instead, move some or all of the lead on the left towards the right — easier said than done. Avoid placing lead high in the chassis and behind the rear end, or forward of the front wheel centerline.

We built a tray on our car that extends from the rear of the left front wheel to the front of the left rear wheel, extending from the frame to the body. This tray is about 8 inches wide and it allows us to move weight fore or aft if desired — the weight ended up as far back and left as we could get it and only 70 or 80 pounds had to be anchored on the right for weight bias. Lead is very bulky and finding places to bolt it to under the little chassis is not an easy task.

Melting your own lead into shapes that can be worked with will make things easier. Since lead fumes are deadly, send the work out to a smelter and give him your mold to melt your lead into. Wash your hands with soap and hot water after handling lead or the stuff might end up in your food, or worse— you.

The Shocks

We used Carrera coil-overs number 5176 in the front and number 5194 for the rear. Initially the shocks were set in the 50/50 position, number 2 on the adjustment.

Position 1 on the adjuster gives the shock 35% extension over 65% compression. This setting allows a faster rate of body lift and a slower rate of compression.

Position 2 is the 50/50 position. This setting is the same as the 3100 series, which are non-adjustable.

Position 3 is 70% extension and 30% compression. This setting allows more extension resistance and less compression force.

The Carrera shocks come equipped with a rubber travel indicator. Move the indicator down to the top of the shock before going out on the track. After you return to the pits measure how much the indicator moved and note the figure. These figures will indicate how much body roll your car has at each end. Recommended figures for shock travel on the Ebeling chassis are: RF — ¾ to 1 ½ inches; LF — ½ to 1 inch, and rear — 2 to 3 inches for both. Note that on a rough track, increased bump travel will change these figures.

The push in the Ebeling Mustang was caused by front end springs being too stiff as indicated here by the lifting of the left front wheel under power.

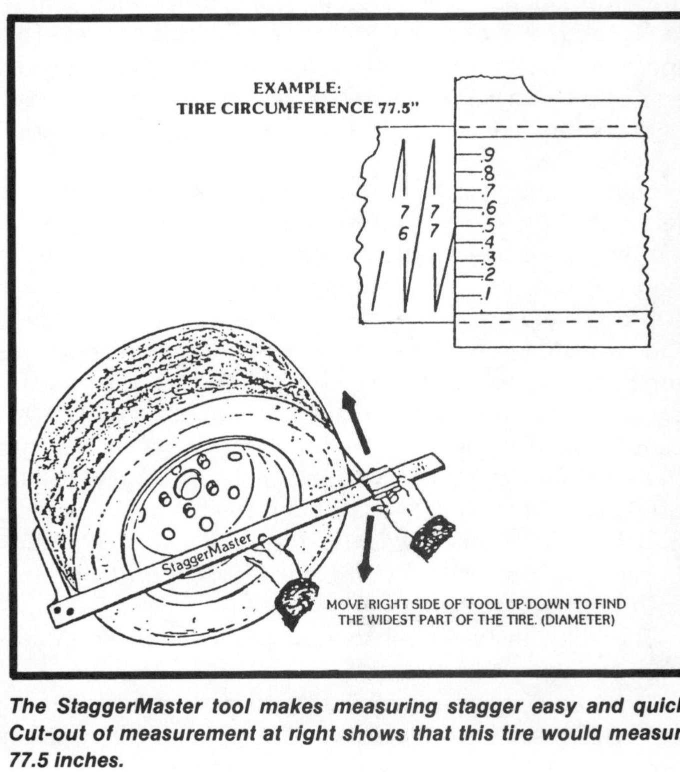

The StaggerMaster tool makes measuring stagger easy and quick. Cut-out of measurement at right shows that this tire would measure 77.5 inches.

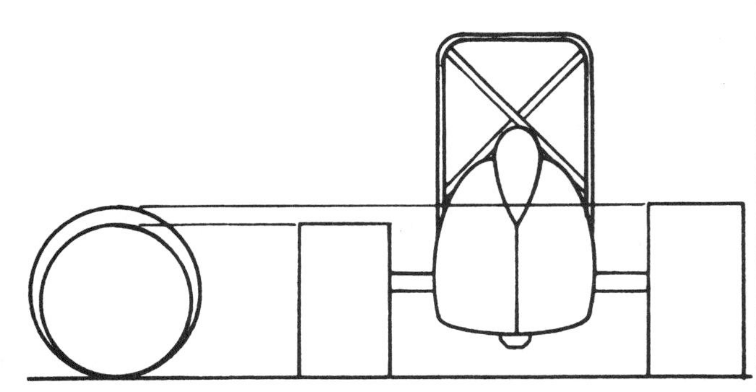

Stagger

Tire Stagger

Stagger is the difference in inches of the tire circumference between the left rear and right rear tires. When the right rear is larger than the left rear, we have stagger. When the left rear is larger in circumference than the right rear, we have reverse stagger.

Short track racing means, for the most part, a solidly locked rear end, instead of the use of a differential. A differential is fine for a street-driven vehicle which has to turn left and right and at relatively very low speeds. The rear end on this street-driven vehicle needs to have the speed of the left and right rear wheels differentiated because their turning radii are very different, and the tires of this car are generating little or no slip angle. But on a race car operating at maximum cornering capability, the slip angles generated at the rear tires negate the need for the differential. Racers use either a spool, which solidly locks the rear wheels together, or a Detroit Locker. The spool-equipped car needs more stagger than the car equipped with a Detroit Locker in the same application, because the Locker unlocks the left rear wheel under deceleration.

Stagger is used as a fine-tuning tool to get the car to turn left in the middle of a turn without having any push in the car. No more stagger than is absolutely necessary should be used to accomplish this. Massive amounts of stagger will only serve to make the chassis excessively sensitive. On asphalt tracks, stagger increases the right rear tire temperature, because of the larger tire wanting to turn the car in a circle about the smaller circumference tire. This causes tire scuff or drag. The more stagger a car has, the more heat the right rear will pick up. More stagger can be used on a dirt track car because dirt track tires run cooler, and in fact, more stagger is required on a dirt track because of the lower coefficient of friction of the tires with the track surface.

The following are some general "rule of thumb" guidelines for using tire stagger:

1) More stagger makes a car tend toward looseness or oversteer. Less stagger — or reverse stagger — makes a car tend toward push or understeer.
2) The tighter the turn radius, the greater the amount of stagger required.
3) A dirt track car needs more stagger for a similar chassis and track size than a paved track car.
4) If the track has multi-groove turns, the driver has to decide if he is more competitive driving the lower or the higher groove. This makes a difference in the amount of stagger to be used. Using the lower groove means the car will be using a tighter turning radius, which requires more tire stagger. This higher groove will not require as tight a turning radius, and thus less stagger is used.
5) The flatter the track, the more stagger required. The more banking a track has, the less stagger needed.
6) On dirt, a wet heavy track requires more stagger. A dry slick track needs little stagger.
7) The wider the rear track width of the race car, the more stagger required for a particular situation. With a wider track width, the right rear is traveling further on the race track than the left rear. As the rear track width of the race car narrows, the right rear is drawn closer to the left rear, and it is traveling a lesser distance, so the circumference of the right rear needs to be a little smaller to compensate.
8) A race car using a wing requires much more stagger, because the wing puts more downforce on the rear driving wheels, especially the left rear. With more forward bite available at the rear, more stagger is used to make the car corner around a turn without pushing.
9) A race car is very sensitive to small changes in stagger. If your car does not react to small stagger changes, there are some other basic problems in the chassis. Don't try to correct them with massive amounts of stagger. You'll only compound your problems. Make sure the basic points in the chassis are in order first — spring rates, alignment, weight distribution, etc. Then, and only then do you use to fine-tune the chassis.

How Much Stagger?

Changing stagger is a much more chassis-sensitive adjustment than changing wedge. In other words, stagger can make a more precise, sensitive change in the car's handling. This is because stagger changes the steering effect of the rear end (stagger makes the right rear want to steer about the left rear) whereas wedge increases the force downward on a tire (no steering effect). A one-inch difference in tire circumference makes a 3/16-inch difference in tire height. So, if you install a tire on the right rear which is two inches in circumference greater, you have raised that corner of the car 3/8-inch (2 times 3/16-inch). That means you have taken 3/8-inch of wedge out of the car. As you can see, the change in stagger has a much more dramatic effect on the car's handling than does the wedge change.

Chapter 9

The Braking System

During the construction of our car we found that working directly with the part manufacturers offered some distinct advantages. The manufacturers have the experience to help the racer design a system suited for his car and form of racing. The manufacturers also have the cumulative knowledge gathered from other racers driving the same kind of cars, and manufacturers have the in-house knowledge and equipment to assist the racer with his design needs. These advantages were evident with the dry sump system, the engine, the chassis, the carburetion, the cooling system and, most importantly, with the brakes.

Sizing The Brakes

The braking system is quite complex and its design should be left to experts, contrary to what most racers believe. Two cars might look identical and race in the same class, but the two cars might have totally different needs for brakes. The parameters that affect brake design are, in part: total weight of the car, frequency of use of the brakes, front to rear weight distribution, form of racing, surface raced on, tires used and their size, center of gravity height, wheelbase, wings or spoilers used, maximum speeds reached, driver's style (some drivers go in the turns with the gas and the brakes on), length of races usually entered, and amount of fuel used in a race.

It is obvious that weighing all these parameters should be left to specialists. Before calling the brake manufacturer, have all this information ready and, if you are updating your present system, or calling about a particular problem, also have all othher specifications ready such as the master cylinder sizes and brand, pedal ratio and caliper size. Rotor size, pad material and the number of pistons in the calipers is also important information to have before calling the brake manufacturer.

For a car weighing around 2,000 pounds racing on oval tracks or road courses, dirt or asphalt, JFZ suggested using four piston calipers front and rear of 1.5 inches bore, and

The stock Pinto calipers and discs are very heavy for use on a race car. The calipers are not rated for racing purposes and should be replaced with units designed for heavy use. Furthermore, steel acts as an insulator and it will cause the fluid to run hot.

10.5 inches vented rotors front and rear. The master cylinders should be 7/8-inch at both ends and the pedal ratio should be 5 to 5.5 to 1. For dirt racing, solid rotors can be used on the front and possibly the rear.

Caliper Rating

Brake calipers on a racing car should be rated at a maximum working pressure of 1,500 PSI, with a minimum burst pressure of 3,000 PSI. Using calipers that are underrated can cause some problems that build on top of each other and that eventually can cause a failure.

The stock Pinto calipers, for example, are cast out of steel, and steel does not dissipate heat as well as aluminum. So when heat builds up in the calipers, the fluid temperature also increases and the pads overheat. Pads can easily lose half their coefficient of friction through heat build-up, thus half their stopping ability.

The driver compensates for the lack of braking by increasing pedal force, which in turn causes the caliper housing to deflect. A driver can apply 100 pounds of force to the pedal. This force multiplied by 6, (the pedal ratio) and by 2 for the hydraulic ratio, becomes 1,200 pounds (100 x 6 x 2 1,200). It takes a well built caliper to handle that force.

The calipers should be rated at at least 1,500 PSI working pressure, with a 2.5 to 3 time safety factor. But a caliper such as the stock unit or some of the low rated aftermarket calipers, or too small a caliper for the application, will be overly stressed and may fail. Stock braking systems are designed to stop a car at relatively lower speeds and at a lower frequency of applications and they should not be used on a race car.

On most asphalt racing cars, the braking force applied at the calipers is around 800 PSI. If the force exceeds 1,400 to 1,500 pounds, the calipers are too small and they should be replaced with units of adequate size. Do not attempt to mask the problem by changing other components.

Another problem that could cause over-pressurization of the calipers is the pad material. Too soft a pad could overheat and fade, or too hard a pad may not get warm enough to bite as it should. Braking system problems should be solved, not covered-up. Installing different pads, higher pedal ratios, proportioning valves or smaller master cylinders will only compound the basic problem.

JFZ has available a pressure gauge, (part number 015-0617), to check line pressure and bias balance. This gauge should be used during hot lapping only, because the gauge causes some pedal sponginess. The gauge is equipped with a valve to shut off the flow of fluid when the unit is not in use. This gauge also comes in handy if the car is raced on different tracks. The setting can be checked, noted and used again at that track the next time the car is raced there.

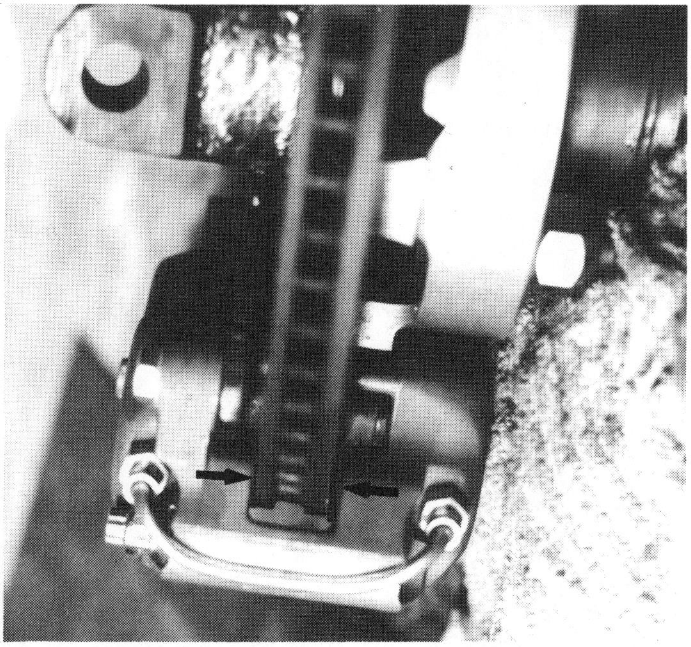

When installing the caliper brackets, the discs must be directly in the center of the caliper or the brake pads will drag. To get an accurate measurement, the brake hat must be bolted solidly to the hub.

The Calipers

The primary function of the calipers is to convert momentum into energy by applying force to the pads and pushing them against the rotor.

The calipers should be machined out of a non-porous casting to prevent holes from surfacing when the bore and o-ring groove are machined. A porous cavity could cause fluid to bleed through the seal. The cavity can also wear the seal and cause a failure. A porous casting is generally weaker than a non-porous one.

The machining tolerances should be maintained to within 0.001-inch. Such a tolerance allows the piston and caliper wall to maintain the exact amount of squeeze on the seal to prevent piston binding and brake pad drag. Incorrect tolerances can also cause the piston to gall the bore, causing a seizure and failure of that caliper. The JFZ calipers chosen for our car have tolerances that are within 0.0002-inch for the bore positioning and diameter. JFZ also uses a dry film lubricant on the pistons to further reduce seal drag.

The machining of the mounting holes is also critical. The bolts that hold the two caliper halves together (the through bolts) should have perfectly aligned holes to thread into, otherwise the bolts may bind, causing stress on the caliper and the pads. This could also cause the bolts to break, the piston to bind in the bore, or the pads to bind in the caliper. Through bolt holes and mounting holes should also be machined to within 0.001- inch tolerance, and the bolts should be a minimum of 3/8- inch, grade 8. Mismatched through bolt holes or inconsistent mounting hole tolerances will at least cause some difficulties during assembly.

Clearance & Mounting Dimensions

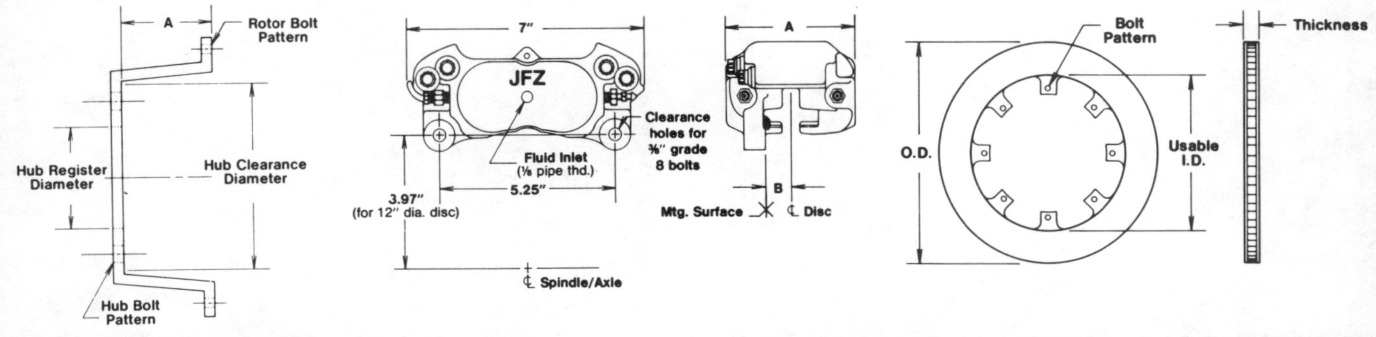

Another item to investigate when buying calipers is the amount of room available around the mounting and through bolts. Some manufacturers do not allow enough room around the bolt heads which makes it difficult to place a wrench or socket on them.

While the weight of a caliper is important because the caliper is part of the unsprung weight, lightening the calipers should not be over-done. Too light a caliper can flex, causing the pads and the piston to bind, or worse, freeze-up in the caliper. Too light a caliper will heat the fluid faster than a heavier one will.

On our car we used the JFZ Mini Grand National dual calipers, part number 002-0347, front and rear. These calipers have 1.5 inches diameter bores and four pistons each.

The fluid seals are an important component of the braking system. The seals operate under extreme pressures and temperatures and only ethylene propelene seals should be used. Only purchase the seals from brake manufacturers to ensure you are getting the right material. A properly designed caliper should have the seal installed in the caliper housing. The reason for this is that as pressure is introduced in the caliper, the seal deforms and it is pushed toward the pad. If the seal is on the piston it will try to keep the piston out when the pressure is released. If the seal is in the caliper housing, it will pull the piston back as it returns to its natural shape. If the piston does not retract fully, the pads will drag, causing unnecessary heat and friction.

The Brake Pads

We used JFZ's Super Pads, part number 005-0300. The Super Pads have the high friction characteristics found in a soft pad and the high resistance to heat normally found in a harder pad.

The following guidelines measure pad characteristics and how they compare to each other: 1) pad compound, which ranges from soft to hard; 2) friction characteristics, which go from high to low; 3) optimum operating temperature, ranging from low to high; 4) application conditions, ranging from less severe to most severe and; 5) fade at high temperature going from greater to lesser. See the chart we included for a better understanding. It is apparent from looking at the chart that a change in one area of the parameters will affect the other areas.

PAD COMPOUND	FRICTION CHARACTERISTICS	OPTIMUM OPERATING TEMPERATURE	APPLICATION CONDITIONS	FADE AT HIGH TEMPERATURE
SOFT	HIGH	LOW	LESS SEVERE	GREATER
↕	↕	↕	↕	↕
HARD	LOW	HIGH	MOST SEVERE	LESSER

Bedding The Pads

The brake pads should be bedded-in to provide maximum life and performance. When replacing the pads, rough-up the discs with coarse sand paper to remove the

These caliper brackets can be obtained from Speedway Engineering and JFZ Brakes. After some grinding they can be mounted to the stock Pinto spindles.

When mounting the brake rotors to the hats or flanges, always use the highest quality fasteners. The brakes get extremely hot and the bolts stretch from the heat. When the bolts stretch they get loose and repeated heat cycles could cause the fasteners to come completely loose. Use the fasteners supplied by the brake manufacturers or aircraft quality fasteners. This rotor is mounted with a grade eight bolt and a locking nut. Use the crushed metal type of locking nut, not the nylon style because the nylon might melt under the extreme temperatures. Also note that the nut is backed up by a thick washer to spread the load evenly and to prevent galling the aluminum.

Lock washers should always be backed up with a flat washer. Here is what happens to an aluminum part if this is not done. Repeated galling will weaken the part.

glaze and to provide a surface for the pad material to seat itself to. There is no need to sand down the pads, or to surface them on a surface plate, because the disc is probably not perfectly flat after it has been used even for one set of pads. Bedding the pads will allow them to take on the same shape as the disc.

Some pad materials need to be bedded with a lot of heat and then left to cool down, while other compounds need more time and more gentle application of the brakes. Consult the manufacturer for his recommendations on bedding his brake pads. The pads are bedded when the scratches from the sand paper have been wiped off the rotor. Be sure you don't bypass this important step in your braking system.

The Rotors

The rotor's job is to take the heat that has been created and dissipate it. The braking system is nothing more than a mechanism that converts force into energy. The force stored in the car (its mass times its speed squared) is converted into energy (heat). The brake pads are pressed against the rotors, creating friction which in turn produces heat. More heat is created as the braking ability of the system improves.

The heat actually does not stop the car — the friction between the rotors and the linings does. Unfortunately the product of friction is heat.

Disc brake rotors are nothing more than levers that are clamped by the brake pads. The larger the diameter of the rotor, the longer that lever arm is and the more stopping power the system will generate. With 13 inch wheels we are limited to using 10.25-inch rotors, which is adequate for our application. The rotors should be the same size front and rear.

The greater the coefficient of friction, or the greater the force or the longer the torque arm, the better the car will stop. However, the better the car stops, the more heat there is to be dissipated. If this heat is not removed from the rotors, it will transmit to the calipers and then to the fluid. This must be avoided.

To remove the heat from the rotors more efficiently, brake engineers have designed a disc is in fact two discs, one for each pad. Since each disc has force applied to it from one side, a method of joining the two rotors to prevent each disc from collapsing had to be devised. Furthermore each rotor, being a lever arm, must be attached to the hub. Instead of utilizing two separate fastening points on the hub for the two rotors, it became more practical to attach one disc to the other via some fins.

It is commonly believed that those fins are used for cooling the rotors by pumping the air through and creating a propeller effect. The major purpose of the fins is to attach the two discs together and to provide some structural rigidity. The fins, in fact, limit the amount of air that can flow in between the two rotors. Staggering the veins helps in getting the air in between the rotors and channeling it to the outside to pick up heat during its stay between the rotor halves.

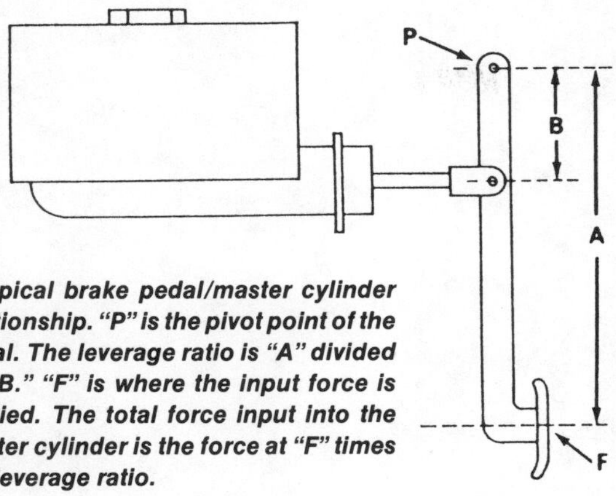

A typical brake pedal/master cylinder relationship. "P" is the pivot point of the pedal. The leverage ratio is "A" divided by "B." "F" is where the input force is applied. The total force input into the master cylinder is the force at "F" times the leverage ratio.

There is another method that is believed to aid cooling. That is the practice of rotor drilling. Drilling rotors is not always a good idea. First, the rotor structural integrity is jeopardized. Remember that the rotors are lever arms and drilling holes in a stressed lever can only weaken its structure.

Furthermore, the rotors get extremely hot and the metal must expand without interruptions. Any holes placed in the path of this expansion will be seen by the metal as stress risers and cracks will take place at the holes.

Of course some rotors can be drilled, such as steel rotors, but the operation should be performed by a knowledgeable operator or by the brake manufacturer who is familiar with the proper sequence of drilling to prevent cracking.

Drilling rotors is done in the hope of creating some venting for the rotors. Just as adding fins only insulates the rotor from ambient air, drilling holes removes the metal that was supposed to conduct the heat to the air. Furthermore material is removed from the discs by drilling, causing loss of friction, which will require greater pedal effort, further heating the rotor.

Drilling rotors does offer one advantage: weight saving. JFZ's solid drilled rotors lose 1.1 pounds during the drilling process and to further lighten the units, JFZ cuts a V shaped groove on the outside of the disc to lighten it by another 1.1 pounds. Drilled rotors are acceptable on vehicles that have a limited need for brakes, such as drag or dirt cars, and where weight is important.

The design and size of the rotors determine how effectively they will dissipate heat. If more heat is generated than can be handled by the rotors, they should be replaced with more efficient units. Drilling the rotors will only compound the problem.

Brake Fluid

Only racing brake fluids must be used in a race car. The brake fluid gets extremely hot under racing conditions and anything but the best fluid will boil and cause a spongy pedal, or worse, the brake pedal will go to the floor.

Cartel brake fluid is rated at 570 degrees and the AP fluid is rated at 550 degrees. These fluids are formulated to withstand racing heat. Regular passenger car fluids are dangerous to use for racing. These fluids will boil at a very low point and they absorb moisture more easily than the racing fluids do.

Silicone brake fluid should never be used in a race car. Silicone fluid has a high boiling point but it becomes compressible at a lower temperature because the silicone fluid throws off a gas. Silicone brake fluids have the advantage of not absorbing moisture and of being harmless to paint, but their other properties make them unacceptable for racing use.

To protect the fluid from moisture, always tighten the bottle caps tightly and do not re-use fluid after it has been bled from the system. Also beware of pressure bleeders that store fluid in the pressure reservoir. The fluid is not air tight in the bleeder and it will pick up moisture.

Checking brake fluid temperature should be one of the items to check when the car is first tested. Use some Tempilaq temperature indicating fluid on the caliper. This fluid comes in different heat ranges and after it is applied to the part, it changes color when it gets to its stated temperature. The outside of the caliper should not be allowed to exceed 500 degrees. Tempilaq kits and individual jars are available from Tilton Engineering.

The Master Cylinders

Again, the master cylinder sizes should be determined by the brake manufacturer. We used Tilton's 7/8-inch master cylinders for the front and the rear brakes.

A few things should be checked on the master cylinders: 1) after the cylinders are installed in the pedal assembly, check that there is no force applied on the piston. The push rod should be perfectly free and not make contact with the piston or residual pressure might build up in the system and cause brake drag; 2) double check that the master cylinder bleed hole (the one nearer the fluid outlet) is not covered by the piston rubber cup, or, again, pressure might build-up in the system and cause the pads to drag or lock; 3) check that the push rods are parallel to each other; 4) the master cylinders should be mounted solidly and higher than the calipers. If the cylinders are lower than the calipers, complete bleeding will be impossible, and; 5) try to install the master cylinders and the reservoirs inside the car. This will prevent the fluid from picking up engine compartment heat.

JFZ manufactures a line of racing master cylinders that feature a removable cartridge element. This cartridge is an independant unit from the master cylinder and it can be replaced quickly when rebuilding or resizing without removing the master cylinder from the car. The JFZ

Remote fluid reservoirs can be positioned anywhere that is convenient as long as they are above the master cylinders. Remote reservoirs facilitate filling the fluid and they leave more room behind the windshield. Remote reservoirs allow the fluid to be checked at a glance.

Rubber brake lines should never be used on a race car. The heat and pressures encountered in race car braking systems surpass the ratings of most rubber brake lines. Rubber brake lines also have the nasty habit of swelling under pressure, causing a spongy brake pedal. These master cylinder push rods should be parallel to each other.

cylinders also feature six large bleed holes, instead of the usual single opening, to allow for rapid fluid return to the cylinder and for quick pad retraction.

The Brake Lines

Rubber brake lines used on passanger cars should never be used on a race car. These lines are not designed to withstand the heat or pressures encountered in racing.

Use -3 or -4 (dash 3 or dash 4) Teflon stainless steel braided lines. The -3 line is more desirable than the -4 line because its smaller size does not allow as much expansion under pressure. Purchase these lines from reliable dealers such as Earl's Supply, JFZ or Baker Precision to ensure you are getting the highest quality lines. Beware of the imported look-alikes.

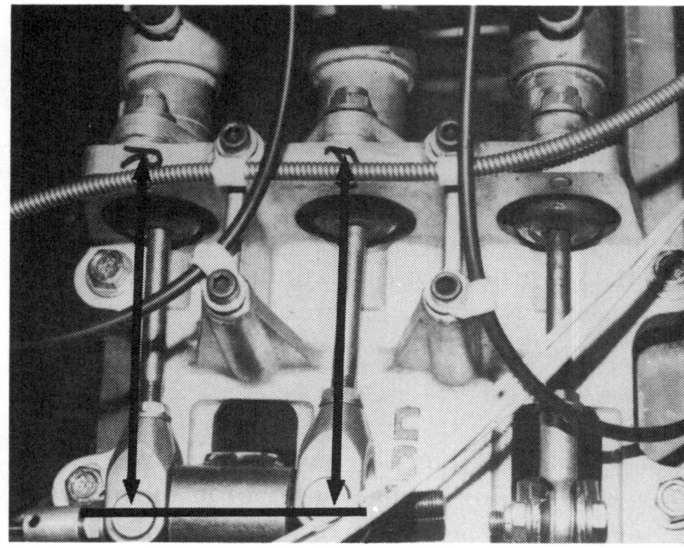

The master cylinder push rods should be parallel to each other. If they are not they will apply force to the cylinder bores causing unnecessary wear. The balance bar should be parallel to the mounting surface of the master cylinders when the brakes are applied. *Arrows show how to measure this distance. Notice how the front and rear cylinders are marked to avoid confusion during repairs and maintenance.*

Before cutting the lines to be installed on the car, use some rope to simulate the hose length and run the wheels through full bump, rebound and, in the front, through full steering lock. Allow a few extra inches of length to prevent crimping of the hose at the fittings. Take extra care to keep the lines away from headers and the brake rotors.

The flex lines should only be used for bridging the steel lines to the calipers. They should not be used instead of steel lines because the braided line tends to deflect under pressure, which could cause a spongy pedal.

Use Bundy Weld Tubing. This tubing is -3 by 0.028 inch wall and it can be purchased from JFZ. Bundy Weld is very strong and it can be bent by hand without collapsing. Cheap passenger car tubing should never be used on a race car. The flaring should be done with a quality tube flaring tool with rollers at the flare. The length of the piece to be flared is also critical, and it should be set with the tool. Add some oil to the part to be flared to prevent galling the tube.

Bleeding

If you are using the dual master cylinder assembly, the front and rear of the brakes must be bled at the same time. If this is not done, the travel of the master cylinder for the end of the car being bled will be restricted by the other master cylinder.

With dual piston calipers, both sides of the caliper

should be bled to ensure all the air has been removed. Using the Tilton brake bleeding bottles makes the job a one man operation. If you have problems getting the fluid to flow to the rear calipers, install a clear hose to the bleeder screw and suck lightly until some fluid can be seen in the hose. Do not allow the reservoirs to run dry, or the whole procedure will have to be repeated.

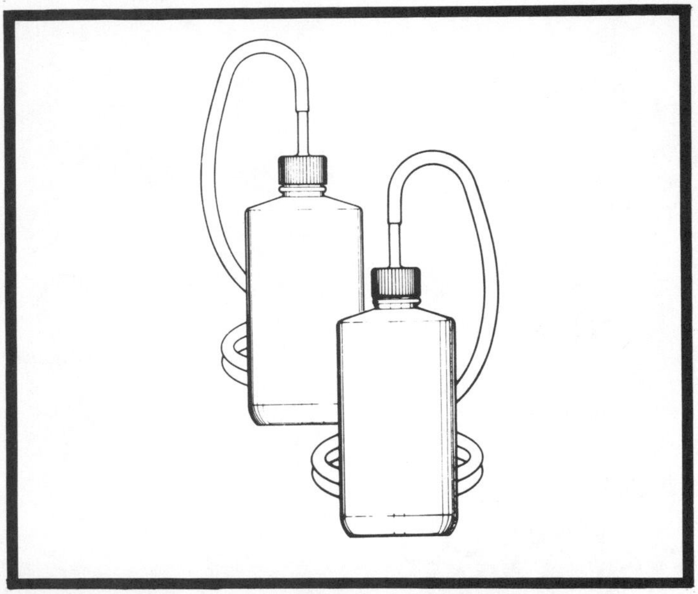

The bleeding bottles available from Tilton Engineering.

Try the wheels on and spin them to check if they touch the calipers anywhere. If a wheel touches a caliper, grind some aluminum off with a rough file. If you need to grind more than 1/16-inch, check the wheels for out-of-round or incorrect offset.

Method of straightening the caliper bracket after welding.

Chapter 10

Wheels and Tires

Bolt Pattern

For Mini Stocks, the most popular pattern is the Pinto and Mustang pattern of four bolts, on 4 1/4-inch spacing. The wide five pattern can also be used, but it doesn't offer any real advantages. JFZ and Speedway Engineering make lightweight hubs for the Pinto bolt pattern, and the stock hubs are readily available. Wheel manufacturers like Bart and Streaker put out some good wheels with the 4 on 4 1/4 pattern.

Wheel Size

Most Mini Stock associations allow 10 to 11-inch wide rims. SCCA limits rim width to 7 inches in its GT3 class and IMSA GTU rules limit the rims to 12 inches. While Mini Stocks and IMSA cars are unrestricted in wheel diameter, SCCA restricts the GT3 class to 13 or 14 inches depending on the body used.

Maintenance and Balancing

The wheels should be visually inspected after every race and Magnafluxed if the car has been in an accident, and always at the end of the season. Cracked wheels can usually be welded, but the reason why the wheel cracked in the first place should be studied and the problem should be corrected.

Bent steel wheels can be straightened if they are not too far gone, but aluminum rims usually have to be replaced. Fortunately, most lightweight rim manufacturers, such as Streaker, use the split rim design, which allows a bent

Streaker makes this 13 inch by 10 inch aluminum spun wheel that weighs 9 pounds. Most steel wheels weigh around 18 pounds.

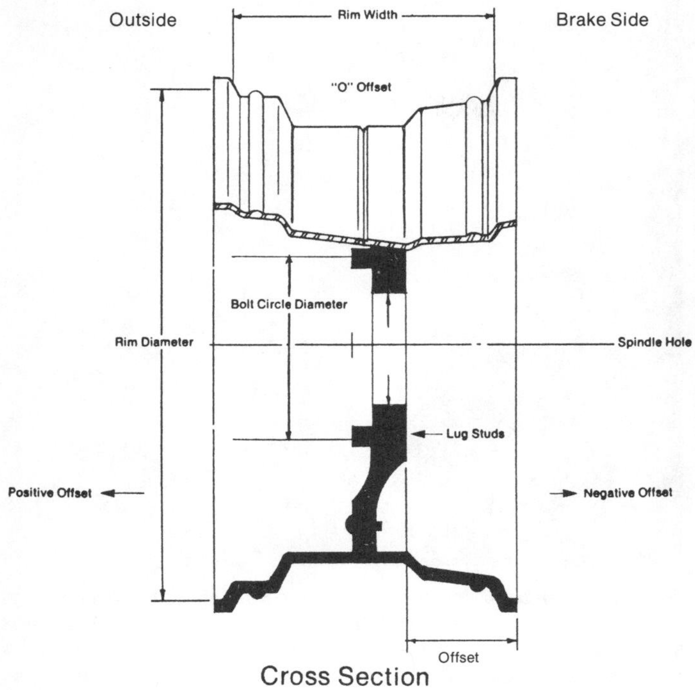

The wheel on the right hit our calipers on the front wheels. The wheel on the left had the rim reversed and it cleared the caliper easily. Bart Wheels can make your wheels either way.

Check the wheel-to-caliper clearance before taking the car out. If the wheel is too close it may rub against the caliper.

wheel to be salvaged by replacing only one half of the rim. In the long run, this advantage might make the split rims more cost effective.

Avoid chroming wheels because the chroming process causes hydrogen embrittlement, which weakens the metal, and the chrome surface can hide cracks.

A newly purchased wheel should be checked for lateral and radial run-out. Lateral run-out is sideways movement, commonly referred to as wobble. Radial run-out is the true roundness of the wheel. Both dimensions should be checked with a dial indicator. The maximum allowable tolerance is 0.020 to 0.025-inch in both lateral and radial.

Wheel Balancing

The wheels should always be balanced, even for short track racing. Speeds can get up to the 70 MPH range even on the smaller oval tracks. An out-of-balance wheel is very annoying going down the straights, and while braking and cornering, the out-of-balance wheel will cause some handling difficulties and discomfort to the driver. An out-of-balance wheel bounces up and down on the track, losing traction momentarily.

A badly out-of-balance wheel can actually get off the track each time the wheel rotates, losing all traction for that moment. When the tire comes back on the ground, it will be overloaded and distorted, and it will again lose traction. For road racing, where speeds reach up to 140 MPH, wheel balancing becomes an absolute necessity. There, an out-of-balance wheel will make the car undriveable at higher speeds.

A wheel balancer is a good investment for a racer. The balancer can be used for the race car wheels and for personal cars as well. Considering the cost of wheel balancing, the tool will pay for itself in short order. Roger Kraus Racing sells a versatile balancer used by most racing teams.

Tire Sizing

The basic tire diameter is determined by three factors: 1)rules, 2)tires and compounds available from the manufacturers and, 3)the size brakes used.

Most Mini Stocks run 13-inch wheels and tires because

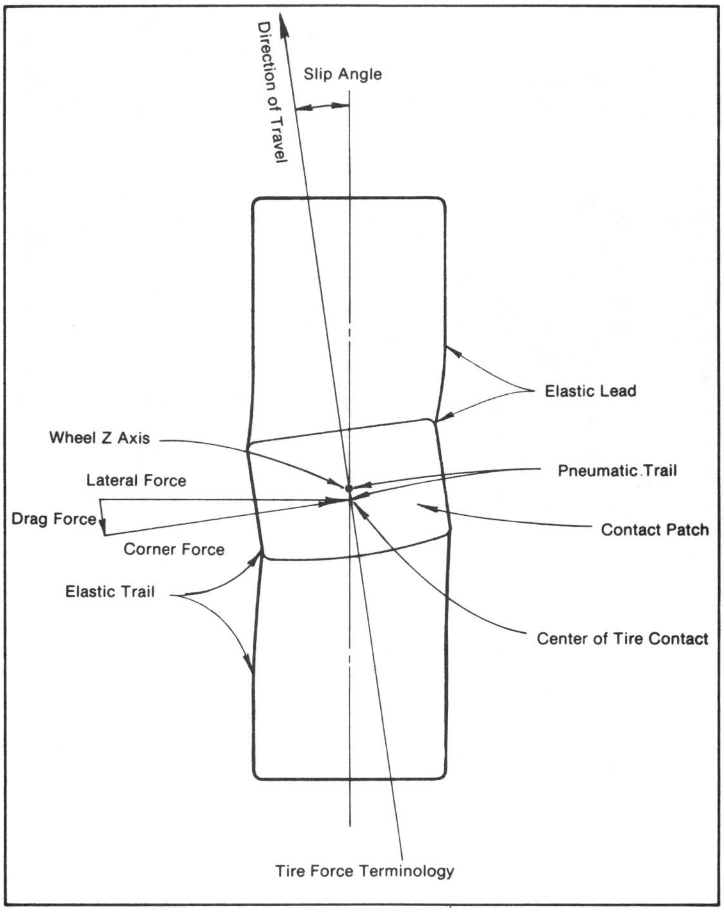

This diagram shows tire forces and their relationships.

This device is used to measure tires. The operator measures the diameter of the tire with the gauge and reads the scale which converts the measurement to circumference. The Stagger Master is more accurate than the old tape measure method, and it does not necessitate jacking the car up.

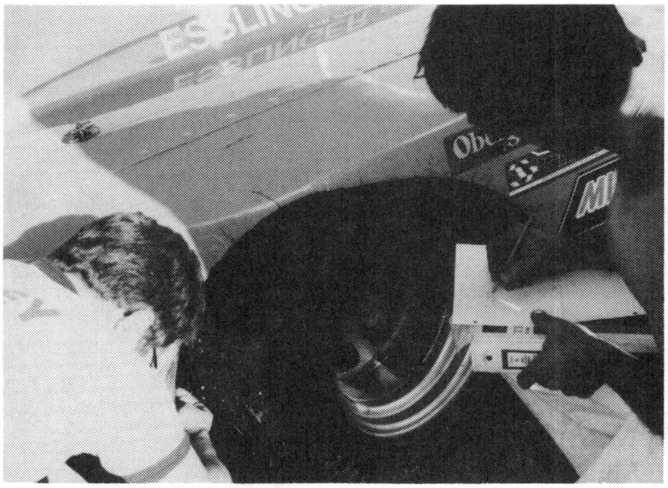

The only way to accurately set up the chassis and the tire pressures is by using a tire pyrometer. We used a Digatron which has a digital readout and is very fast in reading the temperatures.

the tires in those sizes come with the compounds most suitable for short track racing. In road racing it is beneficial to go to the larger diameter wheels to fit the lower profile tires.

The lower profile tire tends to tuck under less and ultimately develops greater cornering forces. Unfortunately for the short track Mini Stocker, the low profile 15, 16 and even 19 inch tires do not take kindly to the camber used in the front end of a left-turn-only car.

Furthermore, in circle track racing the cars use rear end stagger, which induces camber in the rear wheels. The 13-inch tires, with the higher profile ratio, can tolerate as much as 1 degree of camber before losing traction. Meanwhile, the lower profile tires can only tolerate 1/2 to 3/4 degrees of camber before losing traction. The other difficulty of using low profile tires in circle track racing is that the low profile tires are not as forgiving as the 13-inch tires and this makes it difficult to control the car in tight traffic.

Tire Pressures and Temperatures

The only way to accurately determine the optimum tire pressures is by using a tire pyrometer. We used a pyrometer manufactured by Digatron Corp. This instrument is digital in design and it reads temperatures quickly and within one degree. The Digatron pyrometer (and exhaust gas temperature meters) can be obtained from Pitts Performance in Van Nuys, Ca. Tire temperatures should be checked on a skid pad to determine their true temperatures under cornering conditions.

On a race track, the camber, going down the straights, will affect the temperatures. The camber will overheat the inside of the right side tires and the outside of the left side tires, yielding a false reading. The trick is to note the temperatures at the skid pad after the camber and pressure have been adjusted, and compare those with the

These are tire pressure relief valves. These particular valves serve the dual purpose of relief valves and they can also be used to fill the tires. They are available from Edmiston Engineering.

Observe the tire rotation arrow. Tires are designed to work best in one direction. Running the tires in the opposite direction may cause tread separation.

temperatures collected after 8 or 10 hot laps at the track with the same set-up used on the skid pad. By comparing the two results, you can get a rough idea of the effect the straights have on your tire temperatures.

Tire pressures are usually quoted cold. Make a note of the tire temperatures, after running ten to twelve laps on an oval, or three or four laps on a road course. Compare the hot pressures versus the cold pressures. To keep the pressures from building-up, use pressure relief valves such as the ones made by Edmiston Engineering. The Edmiston valves can take the place of the valve stem which eliminates the need for drilling an extra hole in the rim.

For in-depth information on tires and temperatures, read Stock Car Chassis Technology, by Steve Smith.

Tire Compounds

Compounds are the most important feature of a racing tire. Too soft a compound will make the tire stick in the first or second hard lap only to "go away" in the third or fourth lap. Too hard a compound and the tire will not stick until the end of the race, if at all. To choose your tires, go out to the track where you will be running and look at what the front runners use. Tires are in plain sight and that secret cannot be guarded very easily.

Tire compounds are affected by several factors: car's power and weight, track temperature and surface, track configuration, driver's style, race length and rules. Of course dirt track racing will require a different compound and design than an asphalt tire. With our Mustang, we found the McCreary tire to have the advantage of having three different sizes to choose from: 23.5/10-13, a 24.0/10-13 and a 25.0/10-13 tire. The different sizes of tires yield about 1 1/2 to almost 4 inches of stagger at even pressure.

On a 1/3-mile track we found the best stagger setting to be in the 2 to 2 1/2- inch range, running a locked rear end. McCreary compounds go from the softest, number 130, to harder 131, 132 and 133. At our track, the McCrearys work best with the 131 compound and the Hoosiers work best with the A48 compound. Remember that road racing requires a harder compound than for short track racing. Hoosiers do not have the diameter selection the McCrearys have, making stagger setting more difficult. Hoosiers come in 10/19.0-10 and in the A48 "medium" compound and the A52 "firm" compound.

The best tire pressures we found for our car are as follows: right front-18, right rear-19, left rear-13, and left front-13. These pressures were using McCreary tires with the 131 compound.

Both Hoosier and McCreary recommend an optimum temperature of 180 to 200 degrees. Maximum tire temperature is 230 degrees. Be sure to use a tire pyrometer to determine the correct compound for your use, and to sort out proper corner loading each tire.

Tire Break-in

A break-in period is required on all race tires. The reason for this is to scuff the surface of the tire and to remove the residue of the mold release. The normal break- in period is four or five laps on a short track. We want the tires to get to their optimum range and then to cool down. The break-in allows the tire to reach its maximum coefficient of friction. Tire life will also improve by practicing this method.

Chapter 11

Transmission, Clutch and Driveshaft

Which Transmission to Use

I only found four units that are acceptable. The stock Pinto German made four speed, the Borg-Warner T5 or the SROD (short rod overdrive) from Ford Motorsport, and, for all-out oval track racing, the Berthold two speed.

Each unit has its advantages and disadvantages. The stock Pinto unit is light and relatively inexpensive to buy from the wrecking yard, but it is a bit difficult to locate because these units are in high demand by transmission rebuilders. The stock Pinto transmission has worked well behind the 2,000 CC motors, but with the 30 to 50 percent increase in horsepower from the 2,300 with an aluminum head, the long tail unit is marginal. The Mini Stocks using a 2,300 CC powerplant also have to weigh more, creating more stress on the transmission.

The German made four speed should only be used. The English transmission will not stand up to racing stress and parts are no longer available. The German unit can be recognized by the 10 bolt top cover, and the English box by its four bolt cover.

The Ford Motorsport World Class T5 (Ford Motorsport part number M-7003-A) is rated at 305 ft/lbs of torque, well within our limits. This transmission is manufactured by Borg-Warner and the street version is used in the turbo T-Birds and V8 Mustangs, 1984 and later. With a little luck one might find a passenger car version out of a wreck. The World Class version of the street T5 is well suited to road racing because of its five speeds and ease of shifting. The World Class does not need any modifications before going racing, but Tex Powell of Tex Racing Enterprises, who specializes in racing transmission preparation, suggested to use 90 weight gear oil instead of the automatic transmission fluid that Ford recommends for the street. A good racing transmission gear lube is the 75-90W synthetic lube made by Baker Precision. Powell also recommends polishing the sliders and their shaft to prevent binding under power. The teeth should be deburred to relieve stress risers. Neo Seal coating by Baker Precision helps reduce drag, heat and, for road racing, the coating improves durability.

The bellhousing part number for the T5 is: E322-6392-B.

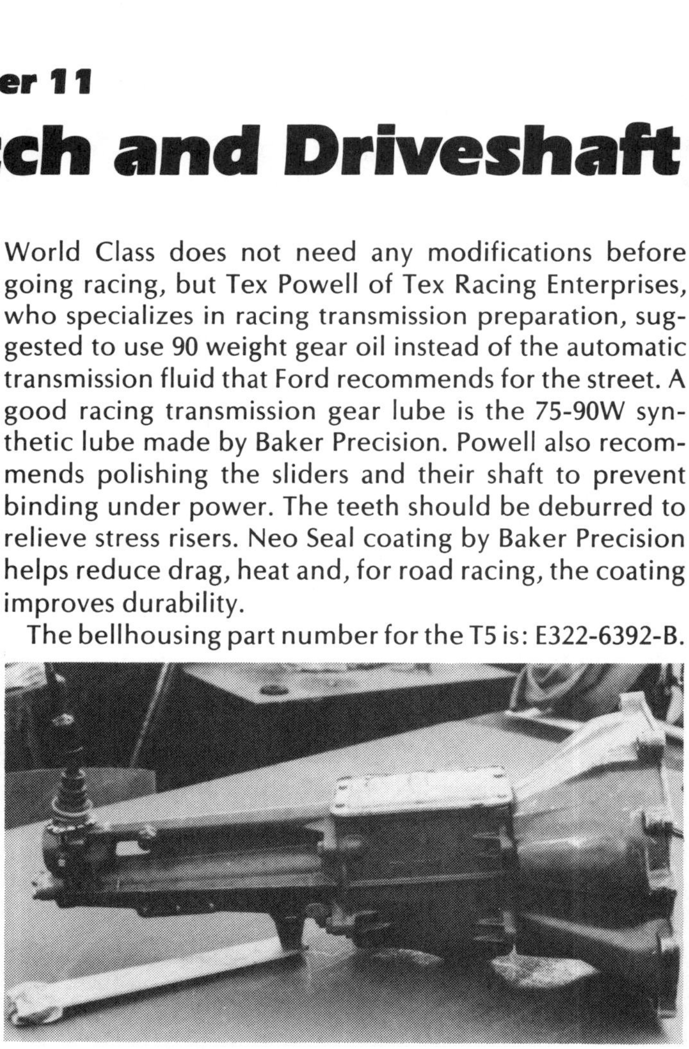

The German Pinto four speed transmission can be recognized by the ten bolt pattern on the top cover. The English four speed has four bolts on the top cover.

The speedometer drive hole should be sealed with a freeze plug and covered with the original drive gear hold-down. We went a step further and safety wired the hold-down bolt.

The four speed bellhousing will not work on the five speed.

The World Class trans from Ford Motorsport differs from the street version in one other area: the input shaft on the World class measures .667-inch, outside diameter, and the passenger car transmission has an input shaft which measures .590-inch. If you run the World Class, you will have to use a larger inside diameter pilot bushing in the crankshaft and allow .007-inch to .010-inch clearance. Ford does not make any brass or roller bushings that fit the 2.3 crank and the World Class input shaft. I found that a brass pilot bushing out of a 1984 Chevrolet S10 will work after the inside diameter is enlarged to accept the T5 input shaft.

The World Class transmission gearing is as follows: 1st—2.95, 2nd—1.94, 3rd—1.34, 4th—1.00, 5th—0.63. The World Class transmission also uses a short-throw shifter. The major drawbacks to the T5 is its high cost and heavy gear mass.

Conversely, the Ford SROD 4 speed overdrive is fairly inexpensive and it has good circle track and road racing potential. It is rated at 263 ft/lbs of torque, and the gear ratios are: 1st—3.07, 2nd—1.72, 3rd—1.00, 4th—0.70. The first and fourth gears can be ground-off to reduce weight and drag. The drawback to this unit for circle track racing is that the gears are rather heavy and cause drag in the oil. Parts for these are also getting scarce. The Ford Motorsport part number for the SROD unit is M-7003-B.

The Berthold Self-Clutching Direct Drive Transmission

The Berthold is intended strictly for oval track racing, where no shifting is necessary. This transmission includes all the other transmissions' advantages at a decent price, considering that the Berthold requires no special flywheel and it comes with its own clutch, pressure plate, release bearing and slave cylinder.

The concept of this transmission is to offer the racer a durable, economical gearbox designed strictly for racing and built from scratch without any stock passenger car parts. The unit weighs about 25 pounds less than a Muncie four speed with the first gear ground off and it offers reduced friction from the lubricant. This weight saving does not include the clutch, pressure plate or flywheel weight.

The case uses the same dimensions as the Muncie and it is cast of 135 aluminum and heat treated to T6 specifications. The case walls are thicker than the Muncie and the mounting ears have been beefed up. The drag caused by the friction of the lube against the gears is dramatically reduced due to the two-speed design and because the gears bathe in automatic transmission fluid. The automatic transmission clutches that connect the engine to the driveshaft are encased in a drum that offers very little drag against the lube. Furthermore, the gears ride on Torrington-caged bearings to reduce friction and heat.

The input shaft connects to a drive flange that bolts to the crankshaft. The aluminum or steel flange comes with 1 3/8 inch-10 Ford truck splining to match the Berthold transmission input shaft splining. Behind the hub is an automatic transmission flex plate to connect the starter to the crankshaft. Berthold also machines their own gears from 8620 steel alloy and they are case hardened 0.030-inch to 0.040-inch deep to a hardness of 60-62 on the Rockwell "C" scale. The gears are considerably lighter than the gears that would come out of a passenger car transmission.

In back-to-back tests on a third-mile track, Berthold experienced an approximate time improvement of 2 to 3 tenths with his unit, compared to a regular four speed.

Here is an easy way to hold the flywheel while torquing it. This tool is made of 1/8 inch plate but any shape scrap will work.

The Clutch and Flywheel

We used the Tilton clutch, pressure plate and flywheel. The clutch cover and plate is part number CP3021 and the whole assembly weighs in at 6 pounds and 6 ounces, with a moment of inertia of 50.5 psi. The "Gray" clutch cover supplies over 600 pounds of breakaway force, plenty for the 2.3. For road racing Tilton recommends the three disc clutch.

The flywheel is steel but it was chosen for its lower moment of inertia, compared to the aluminum unit. The steel flywheel is machined to reduce the outward most rotating weight near the ring gear and to increase the strength near the center.

Try to install the engine and transmission as one unit if you can to prevent this kind of damage to the clutch.

The Release Bearing

We could not get the release lever to fit in the tunnel. The sheet metal would have required some extensive rework and the room for the driver's feet would have been restricted.

The easy way out was to use the Tilton annular clutch release bearing, part number 61-501. These annular release bearings slip directly over the slider tube of the transmission output shaft, and the bearings mount to the bellhousing from the back, where it touches the transmission. The bearings are hydraulically actuated and fully self-contained. They require no release lever or slave cylinder.

Care must be taken to ensure that the bearing sets squarely on its base and that it does not bind on the input

The ARP flywheel Pro bolts are much stronger than the stock bolts. The Pro bolts are available for flex plates and flywheels. Always apply a thin coat of oil to the threads before installing flywheel bolts. Note balancing holes in the flywheel.

shaft. The slider tube must be cut off as the inside diameter of the bearing is smaller than the tube outside dimension.

The bearing must be shimmed to allow a clearance of 0.150-inch between the bearing and the clutch fingers. Shims from 1/16-inch to 1-inch are available from Tilton. With the single disc clutch and the T5 transmission, we had to use almost 1 1/2 inches of shims. Remember that if you go to a two or three disc clutch, you will need less shims. A stop should be installed behind the clutch pedal to prevent damaging the clutch diaphragm springs. This

The Tilton annular release bearing is fully self contained. It requires no slave cylinder or release lever to operate. The line on the left supplies the pressure while the one on top is used for bleeding. Arrow points to the shims that are used to set the bearing to pressure plate clearance.

The Tilton annular release bearing free play can be checked with feeler gauges. Pedal travel should also be set or damage might result to the pressure plate.

can also be obtained by adjusting the clutch master cylinder pushrod length for minimum usable pedal height from the floor.

Bleeding is accomplished via a bleeder kit from Tilton. The bleeder kit is part number 72-510.

Mechanical Linkage

Ford Motorsport has available a heavy duty linkage kit, part number-M-7553-A302. This kit includes a heavier duty release lever and a heavy duty cable. Tilton sells heavy duty mechanical release bearings.

Custom clutch pedals are available from Ebeling Engineering and Tilton. Tilton also has available slave cylinders for hydraulically operated release levers.

The Driveshaft

Ford Motorsport has available driveshaft kits, part numbers M-4602-A. The kits come complete with 55 inches of 3-inch tubing and a 1310 weld yoke. These yokes have 28 splines to fit the Motorsport World Class transmission and they bear part number E2ZZ-4841-A. For the German four speed transmissions, the yoke part number is D1FZ-4841-B and the passenger car T5's use yoke part number E3SZ-4841- A. These yokes have 27 splines. Use 1310 series U-joints from Speedway Engineering.

To measure for the driveshaft, first install the transmission slip yoke in the tailshaft and push it in all the way. Now measure the distance from center of the yoke to the center of the rear end companion flange. The measurement is taken from the center of the U-joint bearing cap location hole, at the transmission and at the companion flange. Take the measurement and deduct two inches from the total. The two inches are to allow the driveshaft to move in and out of the transmission tailshaft as the suspension bumps and rebounds. The splines on the yoke should have a minimum of 3 inches of engagement, or binding may take place while the driveshaft is under power.

The driveshaft can be measured and installed after the ride height and the pinion angle have been set. You will find it much easier to measure and adjust the pinion angle with the body still off the car.

Setting The Pinion Angle

Improper pinion angle setting sets up vibrations in the driveline comparable to the waves created when someone snaps a rope. If a man holds the other end of the rope, he will receive all the energy stored in the waves on the rope. Now, if the second man also snaps the rope at the same rate, both waves will cancel each other. With correct angle phasing we cancel driveshaft vibrations in the same fashion.

We are assuming that the pinion angle is being set on a coil spring, or coil-over, car without any rubber bushings in the trailing arms. To set the pinion angle correctly, we need to know the angle of the engine, the angle of the driveshaft and of the pinion. The difference between the angle of the engine and the angle in the driveshaft must fall within one degree of the difference between the angle of the driveshaft and the pinion. The engine angle is referred to because it is easier to measure the engine angle on the head than measure it at the transmission. For example, let's take a car with one degree of engine angle and a driveshaft at two degrees. The difference in angles is one degree, so the pinion can be set from 0 to minus one degree. Ideally, the pinion angle should be set so that as

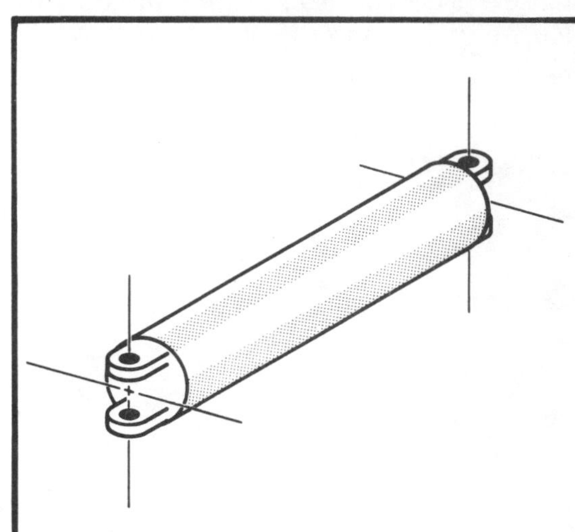

The pivot points of each driveshaft yoke must be absolutely parallel to each other.

the car squats the pinion angle will still be within the acceptable limits we have outlined. To do this the pinion angle would have to be checked with the rear end at the level it would be under full acceleration. This will be a trial and error affair because the driveshaft angle will change each time the pinion is moved. If you want to check the pinion angle, take the angle of the engine and the angle of the driveshaft. Subtract the smaller figure from the larger one. We will call this figure A. Now, take the angle of the driveshaft and the pinion angle and again subtract the smaller figure from the larger one to get figure B. Now subtract A from B or B from A, depending on which figure is larger. The answer should be less than 1 degree or the pinion angle is off.

With a car like ours, with solid motor and suspension mounts and coil springs, Dana engineers recommended to set the rear end parallel to the engine or to set the rear end at the same angle, but in the opposite direction, as the engine. These angles should be measured with chassis in the accelerating mode, because it is under power that the driveshaft is most heavily loaded and that is when vibrations and wear will occur. So, if the engine has one degree of rear sloping angle while the car is under acceleration, the rear end should also be set at 1 degree sloping up or down. The two equal angles will cancel-out the vibrations, just like our two men canceled the vibrations in the rope, by emitting the same vibration from each end.

Forget what you heard about the need for pinion angle. Pinion angle is important on passenger cars. In this case there is a need for some pinion angle to allow the needles to rotate in their cup to stir the grease around and to prevent the same needles from taking all the load, all the time. A racing car dives and squats often enough to allow the angles in the driveline to change and to rotate the needles in the U-joints. Also forget what you heard about driveshaft angle robbing power. Dana Corp. engineers found that even the most out of line driveshaft is still almost 100% efficient, and the amount of power lost is immeasurable.

The Dana Aluminum Driveshaft

The Dana Corporation manufactures aluminum driveshafts for racing cars. These units reduce the rotating weight and reduce vibrations in the drive train. The Dana driveshafts are available from Dana dealers. Dana also puts out a technical brochure on driveshafts and U-joints that is full of valuable information.

The U-joints

There are several do's and don'ts that concern U-joints. First the U-joints must match the companion flange and the yoke. Make sure these measurements are accurate before ordering the parts. Second, Dana recommends the use of a sealed U-joint both for racing and passenger cars. These joints cannot be repacked and they do not have a hole for a zerk fitting. The hole weakens the part and Dana has discontinued the manufacture of these joints because too often the wrong grease was being used, causing a failure. Since the caps do not need to allow grease to escape, as when the bearings are greased, the seals are designed to be perfectly tight, creating better grease retention.

Incidentally, Dana reports that even if the old style U-joints were serviced with the recommended grease and at the recommended intervals, their life expectancy would still be considerably less than the sealed type of joints.

Chapter 12

The Rear End

The Floater Rear End

In the interest of safety, a full floating rear end should be used in any race car. Full floating means the axles transmit torque only, and they do not bear any of the vehicle's weight. If an axle should break in a full floater, the wheel will not come off as would happen in a semi floating rear end as found on passenger cars. The full floating rear end will cost more money to purchase, than a non floater will, but it is an investment that will pay you dividends manyfold in terms of durability and safety.

The floaters can be purchased from Ford Motorsport, part number M-4001-A350 with the 3.50 ratio or, M-4001-A411 with the 4.11 ratio (both 9-inch ring gear rear ends). Speedway Engineering also sells 9- inch floaters with the rear end housing cut to desired length and with the desired spring mounts and locating arm locations. The basic Speedway Engineering floater part number is 7012.

The major drawback to the basic Ford-based floater is that changing ring and pinion ratios is time consuming and ring and pinion sets are costly. In addition, the 9-inch rear end is very heavy, and much more super duty than required for a Mini Stock. There is an 8-inch Ford rear end which is virtually identical to the 9-inch, only smaller and lighter, and much better suited for a Mini Stock. However, the drawback with the 8-inch is that ring gear ratios are extremely limited. This makes the Mini Quick Change floater rear end really the best choice for a Mini Stock.

Housing Width

When ordering a rear end housing from Speedway Engineering, all you have to do is give them the track you want to run and the wheel offset you will be using. The housing flange-to-flange width can then be calculated.

The Quick Change Rear End

The Mini Quick Change rear end from Speedway Engineering includes all the safety advantages the Ford 9-inch floater does, plus a few other advantages that make a quick change rear end so desirable.

The most obvious advantage to the quick change rear end is, as its name implies, the ease and speed with which gear ratios can be changed. All the racer has to do is remove the rear cover, slip off the spacer on each shaft and remove the two spur gears. After the two new gears are installed and the spacers are replaced, the cover is bolted back on and the grease that escaped is replaced. The whole operation can be done by one man in just a few minutes.

The initial cost of the quick change is higher than the Ford-based floater. However, in the long run, the quick change is cheaper because of the lower cost of the gear changes. A 9-inch with three or four sets of gears would about equal the cost of the Mini Quick Change, plus

The Speedway Engineering Mini quick change rear end. These rear ends weigh less than the Ford 9 inch and the gear ratio can be changed in a few minutes. The initial cost of a quick change is a bit more than a nine inch but the added cost will be recovered with the purchase of just one gear change.

changing the ring and pinion is much more time consuming.

The last advantage is in the unsprung weight that is saved. The Mini quick change weighs-in at just under 100 pounds, compared to 173 pounds for the Ford 9-inch floater.

A drawback that is often quoted about the quick change rear end is that the spur gears rob some power in friction. The spur gears do create some extra friction and heat, but the losses in the spur gears are made up by the increased efficiency of the ring and pinion. In the quick change, the pinion sets higher than in the regular floater. This causes less friction between the teeth of the ring and pinion, saving power.

The Spur Gears

The spur gears (also called quick change gears) for the Speedway Engineering Mini Stock quick change are narrower than the big car quick change assembly. The V8 cars need the extra material on the gears because of the added torque placed on the gears. The narrower gears create less friction, saving on power loss and reducing heat.

To calculate the overall gear ratio, divide the number of teeth on the driven gear (upper gear) by the number of teeth on the driving gear (lower gear). Multiply the axle ratio by the answer for the overall rear end ratio. Of course if you run an overdrive transmission or any gear that is not a 1 to 1 ratio you will have to multiply the rear end overall ratio by the transmission gear ratio to compute the overall ratio. For example, using the number 17 gear on the Speedway Engineering chart, the two gears have 29 and 19 teeth. If we place the small gear on the top, the driven side, and the 29 tooth gear on the bottom, we end up with a gear ratio of .65. Multiplying the spur gear ratio by the axle ratio we end up with an overall ratio of 2.74 (19 x 4.11 = 2.74). This ratio is more suited to road racing or a super speedway. If the gears are switched and the 29 tooth is placed on the upper shaft and the 19 tooth is placed on the lower shaft and again we multiply the spur gear ratio by the axle ratio, we now end-up with an overall ratio of 6.27 to 1 (29 x 4.11 = 6.27). As a rule of thumb, for

Speedway Engineering
Gear Ratios V-8 and Midget Rear Ends

3.78 Axle Ratio							4.11 Axle Ratio	
LOW	HIGH	Gear Set No.	Spur Ratio	NO. OF TEETH	Spur Ratio	Gear Set No.	HIGH	LOW
378	378	1	1.00	24/24	1.00	1	411	411
394	362	2	1.04	23/24	0.96	2	394	429
400	357	2A	1.06	17/18	0.94	2A	388	435
411	348	3	1.09	23/25	0.92	3	378	447
422	338	3B	1.12	17/19	0.90	3B	368	459
430	333	3A	1.14	22/25	0.88	3A	362	467
446	319	4	1.18	22/26	0.85	4	348	486
450	318	4A	1.19	16/19	0.84	4A	346	488
468	305	5	1.24	21/26	0.81	5	332	508
477	299	5A	1.26	19/24	0.79	5A	325	519
486	294	6	1.28	21/27	0.78	6	319	528
493	290	24	1.30	20/26	0.77	24	315	536
504	284	25	1.33	18/24	0.75	25	308	548
510	280	7	1.35	20/27	0.74	7	304	554
529	270	8	1.40	20/28	0.72	8	294	575
537	266	22	1.42	19/27	0.70	22	289	584
543	263	9	1.44	16/23	0.70	9	286	591
554	257	10	1.47	15/22	0.68	10	280	603
567	252	11	1.50	18/27	0.67	11	274	617
577	247	12	1.53	19/29	0.65	12	269	627
580	246	13	1.54	15/23	0.65	13	268	630
594	241	14	1.57	14/22	0.64	14	262	646
605	236	15	1.60	15/24	0.63	15	257	658
614	232	16	1.62	16/26	0.62	16	252	668
630	227	17	1.66	18/30	0.60	17	246	685
640	223	18	1.69	13/22	0.59	18	243	696
662	216	18A	1.75	16/28	0.57	18A	235	719
675	211	19	1.78	14/25	0.56	19	230	734
680	210	20	1.80	15/27	0.55	20	228	740
698	205	27	1.84	13/24	0.54	27	223	759
702	204	28	1.86	14/26	0.54	28	221	763
712	201	21	1.88	17/32	0.53	21	218	774
725	197	29	1.92	12/23	0.52	29	214	788
731	196	26	1.93	15/29	0.52	26	213	795
756	189	30	2.00	20/40	0.50	30	205	822
775	184	31	2.05	20/41	0.49	31	200	843
795	180	32	2.10	19/40	0.47	32	195	865

Arrow on the right points to the welding beads used to straighten rear end tubes. The other arrow indicates the Panhard bar mount.

short track racing the smaller gear should go on the bottom shaft and the larger gear on the upper shaft. Another rule of thumb is that changing the overall rear end ratio by .10 to 1 is worth about 125 RPM. If the engine is turning 7,400 RPM at the end of the straight on a short track, and the ratio is changed from a 6.27 to 1 to a 6.46 to 1, the engine will theoretically turn about 7,650 RPM.

Remember to refill the rear cover chamber with gear oil each time the gears are changed, because the grease takes a long time to pass through the pinion bearings into the qc gear chamber. Not filling the quick change gear chamber back up will eventually cause the center section to run low. If the gears are changed several times in a row, check the gear lube level in the center section.

The Spool and the Detroit Locker

There are several ways to lock a rear end. The worst method of locking a rear end is by welding the spider gears. Welding on these heat treated high alloy steel gears can cause stress cracks in the metal, causing a failure under operating conditions.

The best technique of locking a rear end is to use a machined spool to replace the spider gear assembly. A spool will also reduce the rotating mass, especially if it is made of aluminum.

With a Detroit Locker-equipped rear end, any time that overspeed or differentiating must take place, the inside rear wheel is disengaged and it rotates freely. Under straightaway conditions, the Detroit Locker solidly locks both rear wheels. The big advantage of the Detroit Locker for oval track or road racing is that it eliminates inside rear tire scrub found with solidly-locked rear ends.

The only major service problem with a Detroit Locker is that the springs inside it lose their tension after a while. The symptoms of this problem are that the car veers to one side because one axle is not locking up completely. To check the problem, disassemble the Locker and check both the height and rates of the two springs on a valve spring checker.

Only the spool and Detroit Lockers can be used in the Speedway Engineering Mini Quick change. Limited slip differentials and other types of torque sensing devices do not fit in the tight space of the Mini center section. Speedway Engineering sells aluminum spools and Detroit Lockers for the Mini quick change.

Lubrication

Use only the highest quality racing gear oil in a quick change rear end. The quick change has more gears and bearings than the standard floater and the added friction creates more heat.

The following lubricants meet all the racing requirements of both the floater, and quick change rear ends: 1) Baker Precision's Neo synthetic 75-90 gear lube, 2) Pennzoil's 80-90 racing gear lube, 3) Unocal gear lube, 4) Valvoline gear lube, and 5) a 80-140 gear lube available from both Pennzoil and Kendall.

The rear end lubricant temperature should not be allowed to consistently get over 240 degrees F. The rear end lube temperature can be checked by touching the housing with a pyrometer probe, or by applying some Tempilaq temperature indicating fluid on the housing.

For road racing or super speedway racing, a cooler should be used to cool the rear end oil. Tilton sells pumps for pumping the rear end oil through the cooler. The pumps bear part number 17521 and they come equipped with AN-8 fittings. Modine makes a cooler specifically for rear end and transmission cooling. It is part number 1B-109.

Rear End Vents

Rear end housings must be positively vented. Some expansion does take place in he rear end housing and if the pressure is not relieved, it will push the lube through the seals.

Route a 1/4 inch oil resistant line from the plug on the left side bell on the quick change. This plug is higher than the plug on the right bell, which should be used for filling the lube in the rear end, or for checking the level if the breather is used to fill. Install the breather in a location that would allow it to be used as a filler for the rear end.

Chapter 13

The Engine

The 2300 cc Ford OHC motor is the four cylinder of the future at Ford. With the support of Ford SVO and builder/-designers like Dan Esslinger, we can be sure of seeing the birth of the "trick" parts the racers need to compete against adversaries like the Pontiac Iron Duke and the Chrysler 2.2.

The Ford OHC (also dubbed the Lima motor after the city where it is produced) powerplant is already used in the street Mustang, Thunderbird, Merkur and the Ranger pick- up. In the turbocharged form in the Roush Trans Am car, the little 2.3 performed like a champ in its very first races, and it proved its reliability under heavy turbo boost.

Many other motors are used in Mini Stocks: Volkswagen, Toyota, Nissan, Chrysler, Volvo, and Pontiac, among others. There were a few things we wanted in a motor before we started the project. First we wanted to use an American made powerplant. Second we wanted a good balance of performance, reliability, parts availability and economy.

The Ford 2.3 OHC has proved itself in desert racing, drags, stadium racing, boat racing, Mini Stocks and road racing. Parts for the little Ford are as close as your local Ford Motorsport dealer and engine builders like Esslinger or Racer Walsh.

For our reader's sake we also wanted a powerplant built by a company that showed an interest in racing, now and for the future. Ford also offers a nice contingency program through its National Dealers Association, for Ford drivers in all forms of racing (write to: Ford Dealers Association 17000 Southfield Rd., Allen Park, Michigan, 48180).

Choosing The Block

There are three ways to go: 1) find a good low mileage block that will only need a quick hone job on the cylind-

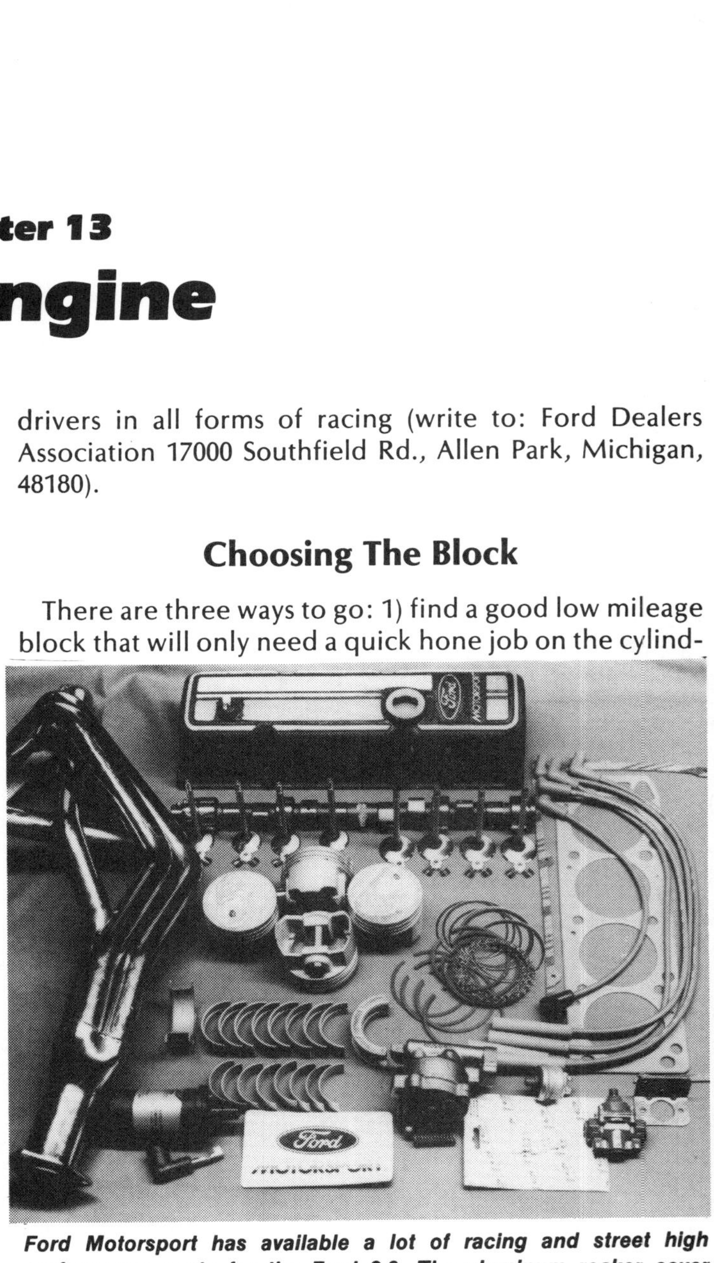

Ford Motorsport has available a lot of racing and street high performance parts for the Ford 2.3. The aluminum rocker cover dresses up a motor very nicely.

ers. 2) If you get a used block with many miles on it you will save a bundle initially, but the whole thing will have to be re-machined. The cylinders and the crankshaft mains will need line-boring, the unit will need hot tanking and the deck will have to be resurfaced. The cost of the machining will almost cover the price of a new block. 3) Buying a new block will save the machine work and the hot tank. The tolerances in the new blocks are close enough straight from the factory. The cylinders will be line honed anyway and the deck height and crank main line bores are going to be checked.

There are a few things to look for when choosing a motor from a wrecking yard: 1) Take a close look at the junction of the breather boss to the block on the left side of the block. This is where a 2.3 cracks if it freezes. 2) Pull the valve cover and, or, the pan to check for cracks, heavy sludge and heavy cylinder wall wear. 3) If you can, pull the head and check it for cracks. With the head off you can check for a cylinder ridge and heavy rust in the water passages.

The heads found on later model Rangers have straight ports and, while these heads are fairly rare, they show some power benefits near the bottom end.

The block and head should be disassembled (remember to remove the freeze plugs and the oil galley threaded plugs), hot tanked (or cleaned with hot soapy water if new), blown dry and sprayed with WD 40.

The head and block should be Magnafluxed to ensure that no cracks exist.

Block Machining

Compared to a V8 or some of the foreign four cylinder motors, very little machining needs to be done to the Ford 2.3 block. On a brand new block only a quick check of the tolerances is necessary. A used block will need to have the deck milled, the cylinders line bored and honed, while the main bearing caps will have to be torqued in place and line bored. The caps should be numbered, from 1 to 5, facing the front of the motor to ensure that all the caps will go back in the motor in the same order they were line bored in. Torque plates must be used while the block is bored and honed. Esslinger finishes the cylinders with a #600 Sunnen hone.

Next the deck is grooved to accept the O-rings. The O-rings are made of 0.041-inch stainless steel wire cut to a zero gap. The important thing to remember about the O-ring procedure is that the rings should rest around the outside of the fire ring to help back it up. If the O- ring is installed directly on the fire rings in the gasket, the gasket will blow in short order. Dan Esslinger reports that this is a common mistake made by amateur engine builders.

The block will need some grinding at the base of the cylinders, near where the rod bolts pass during their rotation, if custom rods are used. The stock rods do not need any more block clearance (unless they are used in conjunction with a stroker crankshaft) but Esslinger still grinds the blocks even when they install stock rods in the motor. The reason for doing this is in case the engine owner decides to install aftermarket rods (a procedure commonly performed by customers because of its apparent ease). Without grinding the block, the rods will collide with the block.

The base of the cylinders is also worked on. The base is chamfered to reduce wear on the piston skirt and to improve piston oiling.

The rear main oil hole is enlarged to allow more oil to pass to the bearings. This is not a problem area, but the procedure is just good insurance. Another modification done by Esslinger is the pinning of the center main bear-

The main bearing oiling holes are grooved to increase the flow of oil to the main and rod bearings.

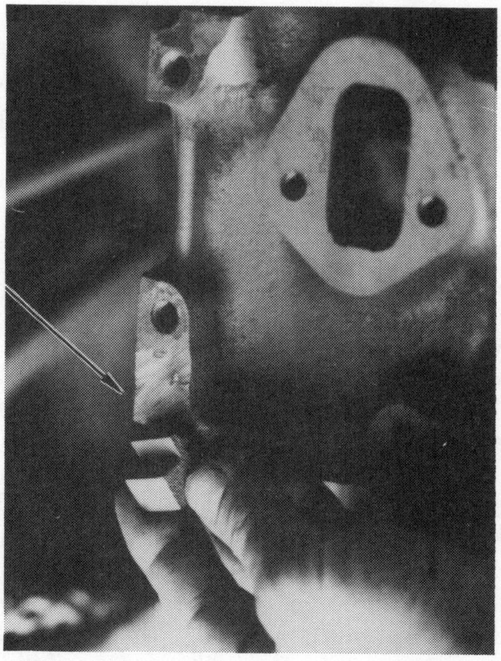

This part of the block must be sawed off if you are going to use the mounts we described in this book. Use a gasket where the stock fuel pump mounted.

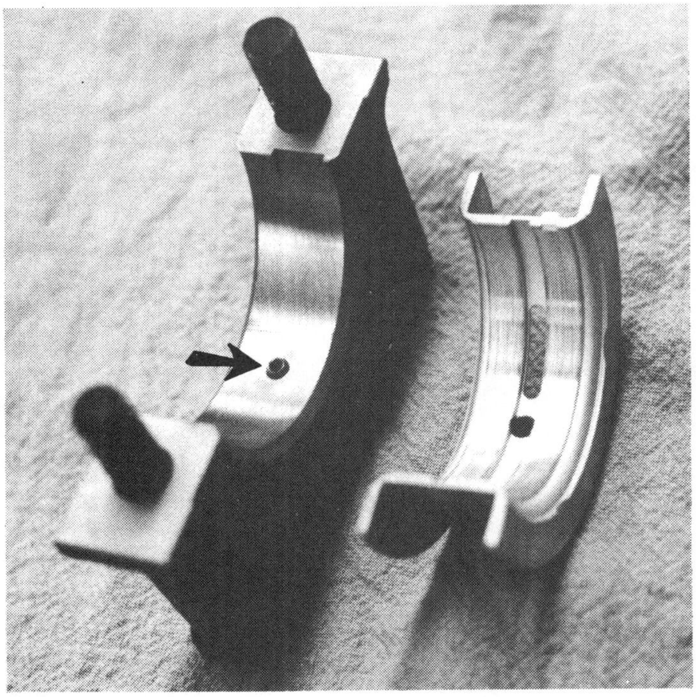

Esslinger pins all the center main bearing caps. To accommodate the pin in the bearing cap, a hole is drilled in the bearing shell. Note the slot in the bearing oil groove. Another photo in this chapter shows how Esslinger grooves the bearing saddles to allow oil to flow to these slots. Oiling is very critical on the Ford 2.3 after 8000 RPM.

ing. Again, this is not a common area of failure, but problems have occurred with spun center mains. The pin is installed in the block side of the bearing bore and the bearing is drilled and relieved to accept the pin. Only the center main is pinned, because it also handles the crankshaft end thrust.

Drilling and tapping the oil passages at the oil filter mount will allow the installation of AN fittings, saving the cost of a remote pick-up adapter for the block. These adapters have a tendency to leak. Steel fittings should be used in the block to prevent galling and leakage problems.

Always install new freeze plugs and reinstall the oil plugs with Teflon tape.

The Oil Pump and Oil Pan

Unless you run a dry sump system, the best bet is a reworked Esslinger pump. This pump is basically a Motorsport unit with some massaging by Esslinger. The volume is left alone, but the pressure is increased, along with some other proprietary tweaks. If only the pressure is improved and the volume remains intact, less oil flows through the by-pass robbing less power and causing less heat in the oil.

Only an oil pan designed for racing should be used on a racing engine. If the oil pump picks up some air in a turn or during acceleration or braking, the motor will starve momentarily and it will eventually seize. The Esslinger pan is designed with several gates and baffles to keep the oil safely around the pump pick-up.

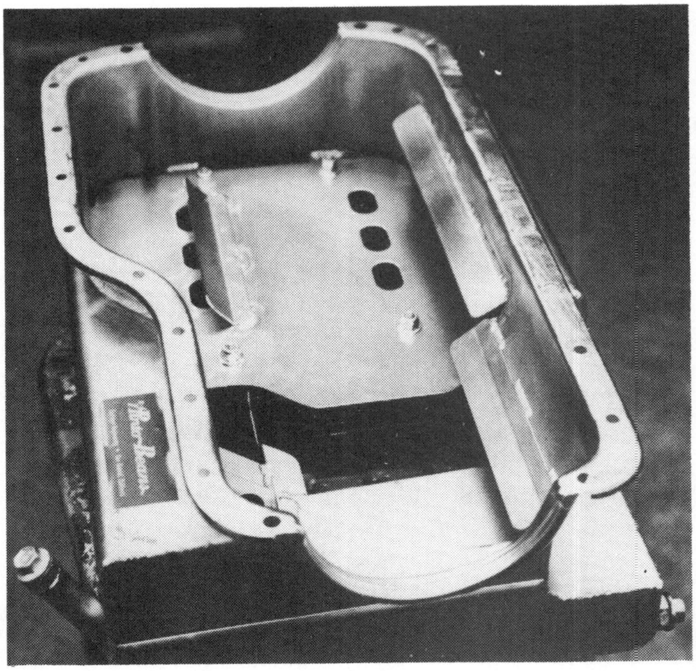

The Esslinger wet sump pans are the next best thing to a dry sump system. These pans use trap doors around the pick-up area and baffles and scrapers to reduce windage around the crankshaft. This pan is designed to insure a constant oil supply to the pump, and also to reduce power losses due to windage.

This oil pick-up should be used with the Esslinger wet sump pan. This pick-up is designed for oval track and road racing.

The Crankshaft

The stock crankshaft has worked well under the most severe conditions, but a few modifications have improved the reliability even more.

Before any work is performed on the crankshaft, it must be checked with micrometers, to see if it is out-of-round and for wear.

Cross drilling the center main on the crank improves oiling to both the mains and the rod bearings. This proce-

Dan Esslinger cross drills all the 2.3 motor crankshafts to improve oiling to the rod bearings

Our crankshaft after it was cross drilled, polished, balanced, filleted and lightened by Esslinger. In the background our Crower rods await installation.

dure is tricky and some specially designed equipment is necessary to perform the operation. Only a professional engine machinist should attempt this. The oil holes in the crank are chamfered to relieve stress risers. Once the machining is completed, the crankshaft is balanced and the bearing surfaces are micro polished. The crankshaft end-play is set at 0.004 to 0.007 inch.

For special applications, Esslinger Engineering offers a stroker crank of their own design with a stroke from zero to 11/16th-inch over stock. Combined with a slight overbore, the longer stroke brings the total displacement up to 2,900 CCs. These cranks are of better quality than the stock unit and they offer the added advantage of having rough bearing surfaces that can be machined to racing specifications. A smooth filet is machined in the rod and main bearing throws.

These cranks are used on some Mini Stocks for open comp, or on motors with sleeved cylinders — the sleeves

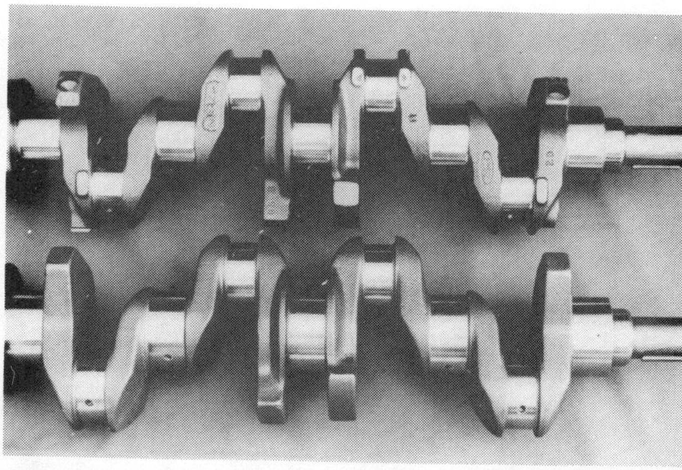

Bottom part of this photo shows the Esslinger stroker crankshaft before polishing. The Esslinger crank is lighter and stronger than the stock unit.

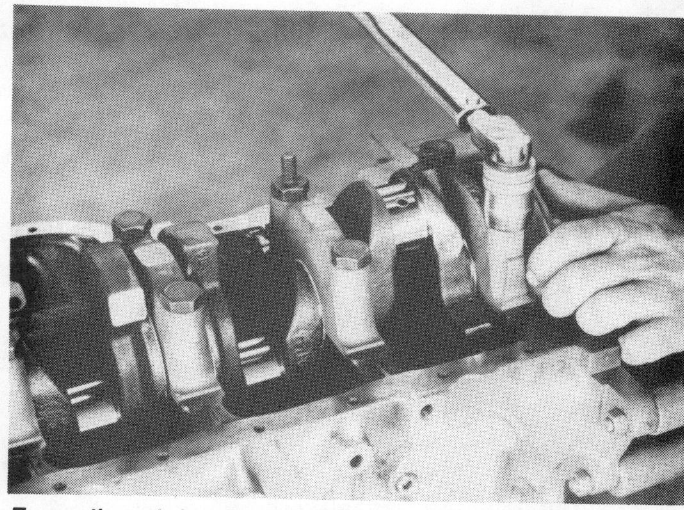

Torque the main bearings to 100 foot pounds. Esslinger likes to torque one cap at a time and then he rotates the crank. This checks each individual bearing cap for high spots. Note the stud sticking out of the number three bearing cap bolt. This stud holds the oil pick up.

Set the crankshaft end play between .004 and .007-inch. End play is read off a dial indicator and it is taken by inserting a screwdriver on each side of one of the crankshaft flanges.

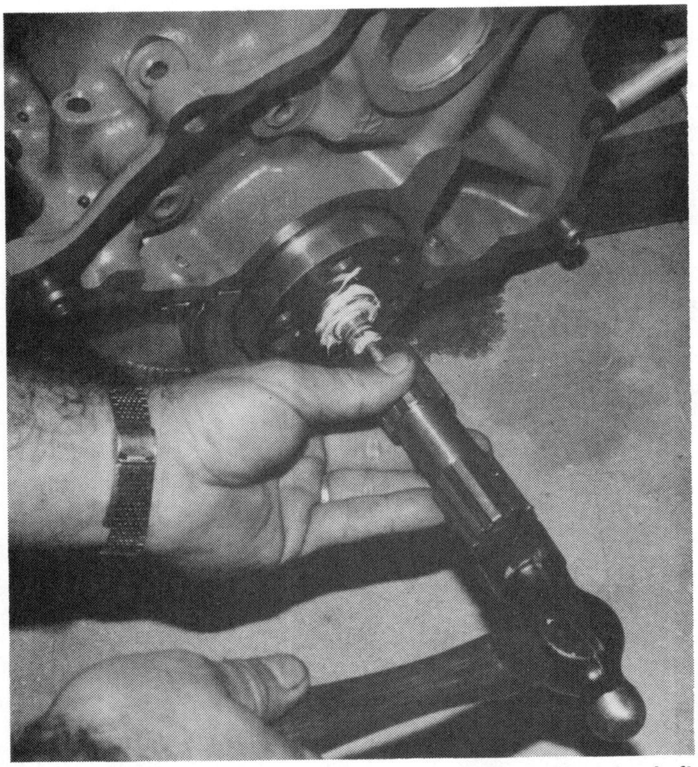

To remove the old pilot bearing, pack it with grease and insert a shaft about the size of the transmission input shaft. Tapping the shaft with a hammer will build pressure behind the bearing that will push it out.

reduce the bore, while the crank increases the stroke to bring the engine size back up to the desired displacement. The idea behind this is to increase the bottom end torque by increasing the stroke. Before installing the crankshaft, apply a liberal coat of white bearing grease to the bearing surfaces. The main bearing caps are installed in the same order they were removed, and are torqued to 100 foot-pounds. The crankshaft end paly is set at 0.004 to 0.007-inch.

Connecting Rods

The stock rods are reliable in mild motors, but after 7,200 to 7,500 RPM or over 175 horsepower, the stock connecting rods become marginal. The other drawback to the stock rods is that the pin is press fit in the small end. This makes assembly somewhat more difficult and some power is lost in friction. To handle the load, Esslinger turns to Crower steel rods, or their own titanium rods.

The Crower steel rods weigh 624 grams compared to the stock units at 635. The aftermarket rods are not much lighter, but they are reliable to any RPM the 2.3 puts out, even in turbo motors. The Esslinger titanium rods weigh a mere 410 grams and they are available in 5.5 and 6- inch lengths, versus the 5.2 inches of the stock rod. The cost of titanium connecting rods is high, but the reliability is well worth the extra expense.

Modifying the stock connecting rods requires a lot of time and work. First the rods are Magnafluxed to detect any possible cracks. If the rods pass, the big ends are miked to check for any possible out-of-roundness and to accomplish the right amount of bearing crush. Then the beams are ground on a belt sander to remove any excess metal and stress risers. The top of the small end is ground on the sander until the whole top is smooth and evenly shaped. The sanding motion on the beam should be done along its length. At this stage, the rods are inspected to see if any of the small imperfections remain. The rods are then polished, shot peened and balanced.

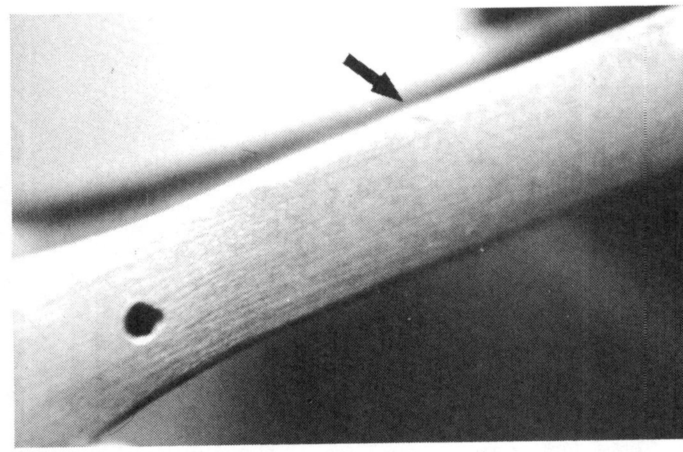

Stock rods must be inspected for casting flaws before being used in a race motor. Esslinger discarded this rod because the casting flaw did not come out during grinding and polishing.

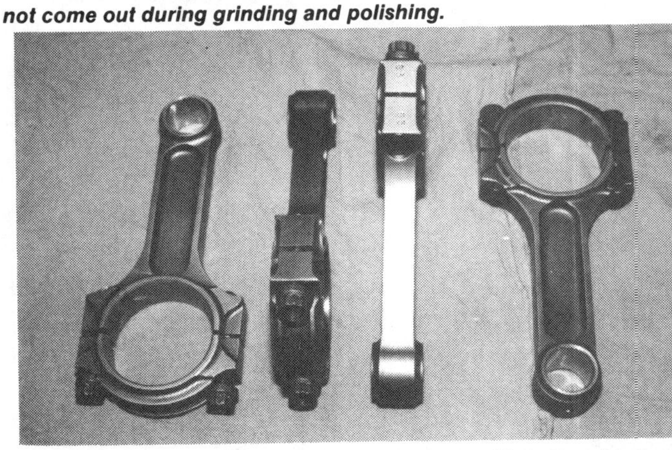

Crower's Pinto steel rods are used by Esslinger. Note the wide beam and flawless forging. These rods are good for any RPM the 2.3 can dish out.

All the Esslinger stock connecting rods are fitted with bolts from A.R.P. The stock bolts should never be used for racing.

When installing the rods and pistons, slip a couple of pieces of rubber tubing over the rod bolts to prevent damage to the crankshaft bearing races. Each rod is installed individually and the crankshaft is rotated to ensure that each rod is turning freely. The connecting rod bolts are torqued to 65 ft/lbs., and connecting rod side clearance is kept at 0.012 to 0.015-inch.

Once the pistons are installed, the deck height is checked to yield the compression ratio desired.

The connecting rod side clearance must always be checked. This is an area where oil pressure can be lost by too wide a clearance.

The deck height is checked using this tool. On our motor with the TRW pistons, the deck height was set at zero. The dome height is also checked to prevent contact with the valves.

Bearings

The rod and main bearings are first roughed-up with 400 grit wet-or-dry sandpaper dipped in solvent. This will remove any rough edges and leave a surface for the oil to cling to. Bearings are TRW Clevite and they run with 0.003 inch of clearance if the engine is to be broken-in on the dyno, or 0.0035-inch if the engine is to be broken-in on the track.

Esslinger does not recommend using bearings over 0.020- inch undersize, or heat dissipation may become a problem.

The Pistons and Wrist Pins

Esslinger generally uses pistons from TRW and Ford

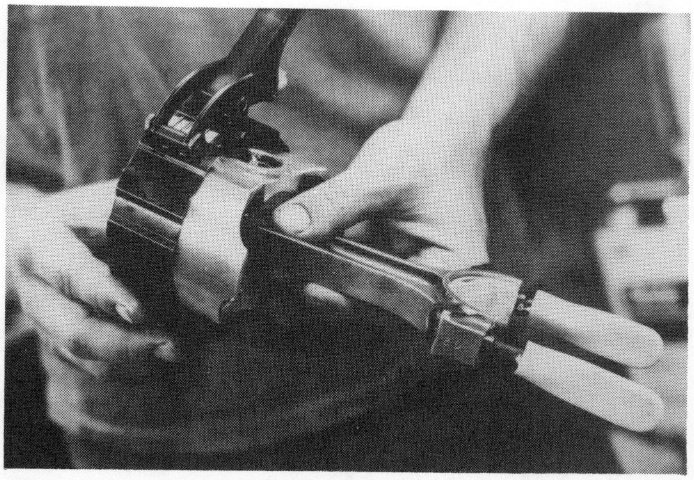

Always cover the rod bolts before installing the piston to prevent damage to the crankshaft bearing surface. These rubber bolt covers can be purchased from any speed shop. Note the ring compressor. It is corrugated to squeeze the rings and the outside of the compression strap seats against the deck of the block. Crower rods come already numbered to facilitate assembly.

Motorsport, depending on the application. For an all-out motor, or if customer wishes a custom ultra lightweight piston, Esslinger uses BRC as a supplier. BRC makes only custom pistons and the cost is high. But for a custom motor these pistons are the way to go.

For the 2.3 Ford, Esslinger prefers to mill the head rather than running a domed piston to increase the compression. This seems to work better for most applications. The Esslinger motors, with the aluminum head, seem to perform better with the TRW pistons with the small dome.

The BRC custom pistons built for our project motor were about 70 grams lighter than the TRW pistons, and they were pre-balanced.

Our motor was fitted with Esslinger ultra light wrist pins. Each wrist pin is retained by four snap rings.

These pistons were custom made by BRC. BRC makes custom designed pistons for Esslinger Engineering for special applications and for research. Note the light-weight wrist pins.

Piston Rings

We chose Total Seal gapless rings for our pistons. These rings have the unique design that uses a thin compression ring and two thinner spacers. One spacer fits on top of the ring and the other fits on the bottom to completely seal the ring gap. These rings have a very good sealing ability, and Esslinger has found good success with them. The Total Seals also tend to reduce oil temperature because of the lower blow-by, and oil dilution is also reduced because of the rings' sealing qualities. Since less hot gasses pass by the top ring, all the rings run cooler and last longer. Ring end gap is set at 0.012 - inch.

After the rods have been torqued, set the pistons half way through their stroke and do not move the crankshaft until ready to time the camshaft.

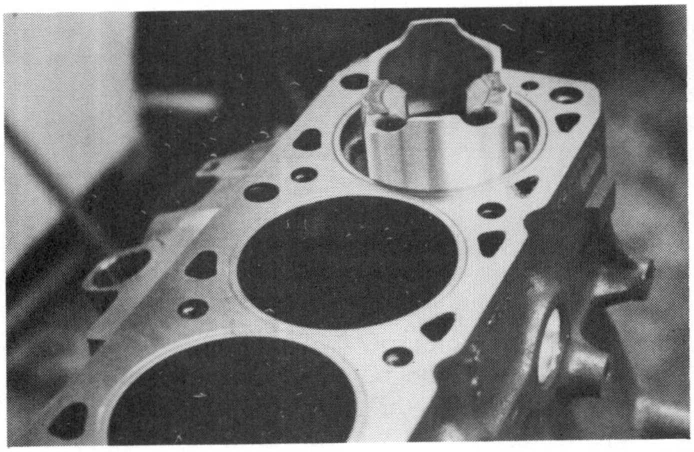

To measure piston ring gap, insert the ring in the bore and push it down with a piston to get it square in the bore. Note the o-rings around the cylinders. These should be positioned outside the gasket sealing ring area.

The Cylinder Head

The stock cylinder head has many attributes that make it desirable to the budget racer. They are easily obtained from wrecking yards at a decent price and new parts are readily available from Ford dealers. The basic crossflow design is good and it can flow well, after some porting and increased valving. The only drawbacks are in the intake port height and angle of the number one and four ports. The last two ports, at each end of the head, are angled out toward the extremities of the head, and they start at the manifold mounting face as huge oval shapes that really hurt the flow and power.

To compensate for the port design, Esslinger has developed their own proprietary port grind. Number one and four ports used to get welded-up to allow re-grinding. After Esslinger Engineering started flow bench work, they discovered that this method really did nothing in terms of flow and an unwelded port head was dynoed with the same results as the welded models. Doing away with the welding saved the customer a lot of dollars for the same amount of power output.

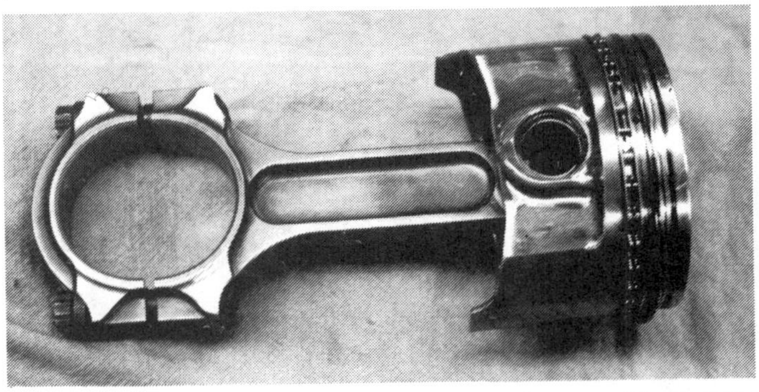

One of our Crower rods and TRW pistons ready for assembly. Note the double wrist pin snap ring arrangement.

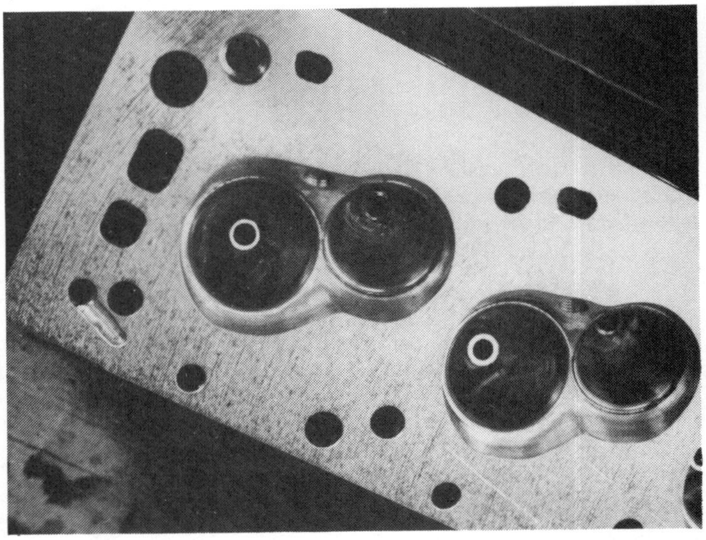

The Esslinger head at the top displays a different shape combustion chamber and spark plug positioning. Also notice the hand grooved channel on the aluminum head to help oiling to the camshaft

Prepping the Stock Head

After the head has been hot tanked and Magnafluxed, it starts undergoing the porting process. This process is very time consuming and should not be attempted by the amateur engine builder. The wrong port shape can yield worse power figures than with no porting at all.

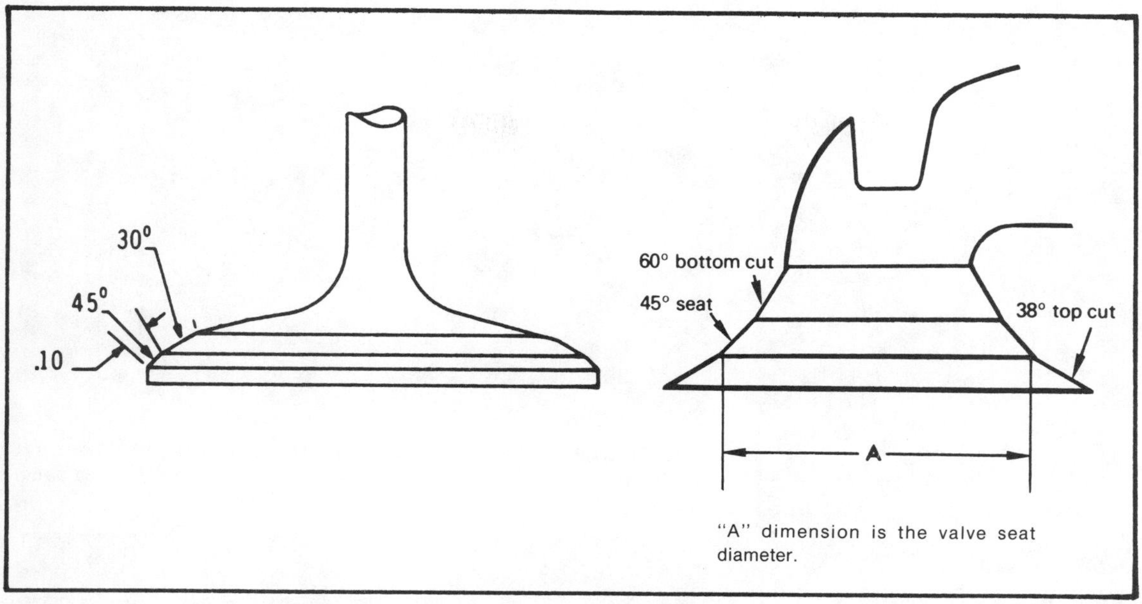

Above left, a typical three-angle high performance valve cut is shown. Above right, the three-angle head seat cuts are shown. At right are all of the terms associated with high performance head and valve work.

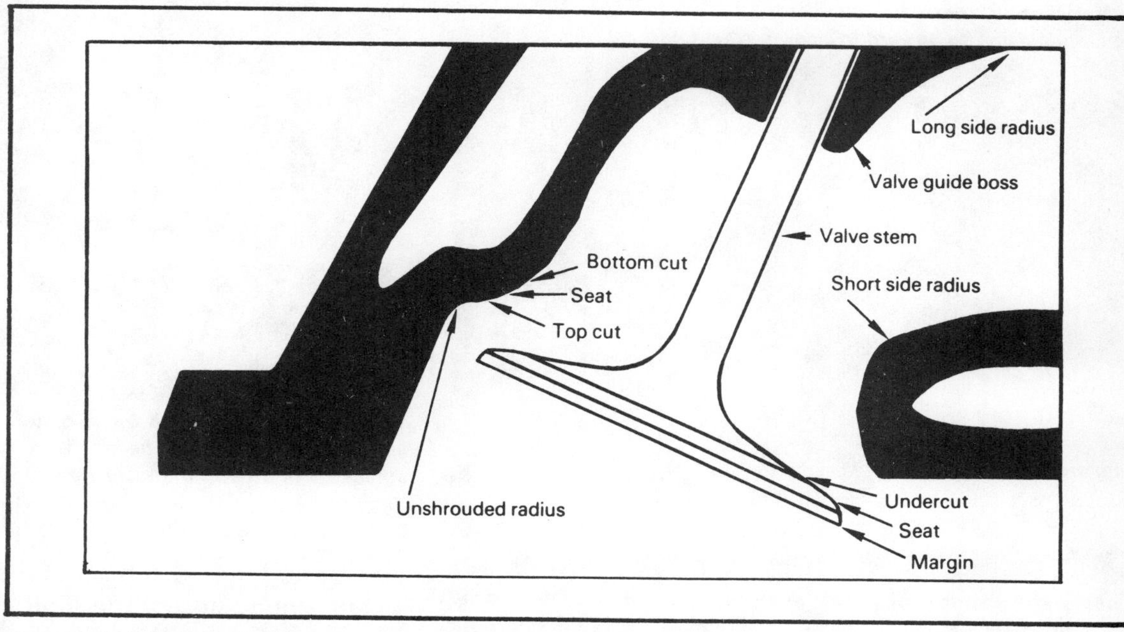

The cylinder head flows the following amounts:

	Intake	Exhaust
Stock head and valves	150 CFM	109 CFM
Modified head, big valves	209 CFM	159 CFM
Esslinger head, as cast	251 CFM	209 CFM

Once the intake and exhaust ports are reworked, the combustion chamber is shaped to Esslinger's specifications and CC'ed to ensure an even compression ratio across all four cylinders.

The stock 2,300 head comes with hydraulic lifters. These are not acceptable for racing and they must be replaced. Esslinger uses the 2,000 CC motor rocker studs to support the rocker arms. To install these, the head must be machined to remove the stock lifter bores, drilled and tapped. The spark plug holes are machined to accept flat seat plugs, with washer.

Once all the machining and porting is completed the valve job can begin.

The Esslinger Aluminum Head

In order to compete with the Volkswagens which had enjoyed several decades of research and development, the Iron Dukes that received the blessing of General Motors and the foreign onslaught from Nissan, Toyota and Mazda, Ford had to fight fire with fire and design a cylinder head that was capable of out-performing the establishment. Dan Esslinger went to work on the project and came-up with a design that has proven itself in every form of racing from Trans Am to Mini Stocks and off road racing.

The Esslinger designed and produced head flows con-

Esslinger aluminum head and intake manifold, Weber carburetors and K&N air filters. Under the manifold is the System 1 oil filter. Notice how the manifold slopes upward to clear the distributor.

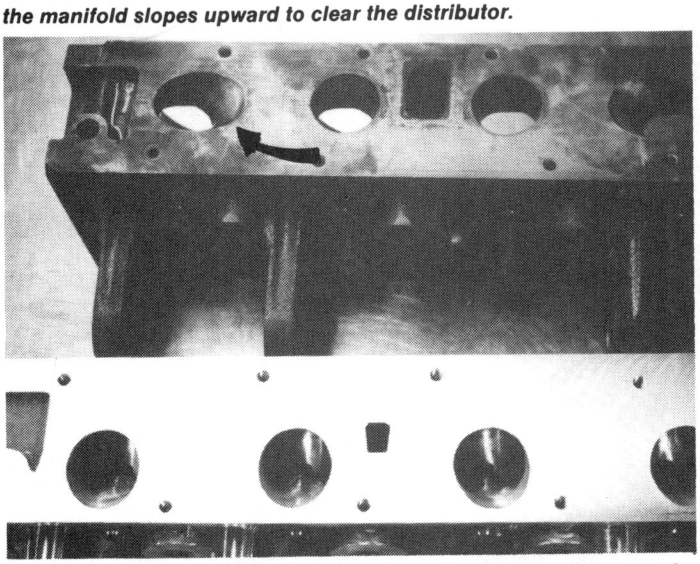

Arrow points to odd shape of the numbers 1 and 4 intake ports on the stock head. The Esslinger head, at bottom, displays straight high ports. The aluminum head weighs 25 pounds less than the stock head.

siderably better, even in its raw casting form than a ported cast iron head. The intake ports are a shape that is conducive to improved gas flow and they are placed high in the head to provide a straight shot to the incoming gases.

The combustion chamber is of the high swirl design and it is shaped to facilitate gas movement, in and out.

The result is a head that increases the power output by some 30 to 50 horsepower, depending on the application. The aluminum head also cools better and it saves a whopping 25 pounds of high center of gravity weight. Because Dan Esslinger also designed the head to accept 12mm spark plugs, this leaves more room in the combustion area for the valves and more metal around the intake port for cooling.

Cost-wise, the aluminum head is not much more expen-

The Esslinger aluminum head is the best power improvement per dollar spent on the Ford 2.3 motor. Here we see the bronze valve guides and the mechanical valve adjuster post holes.

TRW or Ford Motorsport valve guide seals are used on both the aluminum and the stock heads. Note the large intake ports and the steel washer around the valve stem at the center. The steel washers must be used on the aluminum heads to prevent the springs from cutting into the aluminum.

sive, considering that most of the machining has already been done to accept the solid lifter studs, much of the porting is done by the foundry and the valve guides are installed. On a horsepower to dollar basis, the Esslinger aluminum head is the best buy.

In the Ford Motorsports catalog the aluminum head bears part number, M-6049-A230 and it is available both from Ford Motorsports dealers and Esslinger Engineering.

Head Torque

Both the steel and the aluminum heads should be torqued to 90 ft/lbs. An important part of this procedure is to run the engine for an hour or two and re-torque the head after it has cooled.

Valves

Esslinger uses two suppliers for valves. For the steel

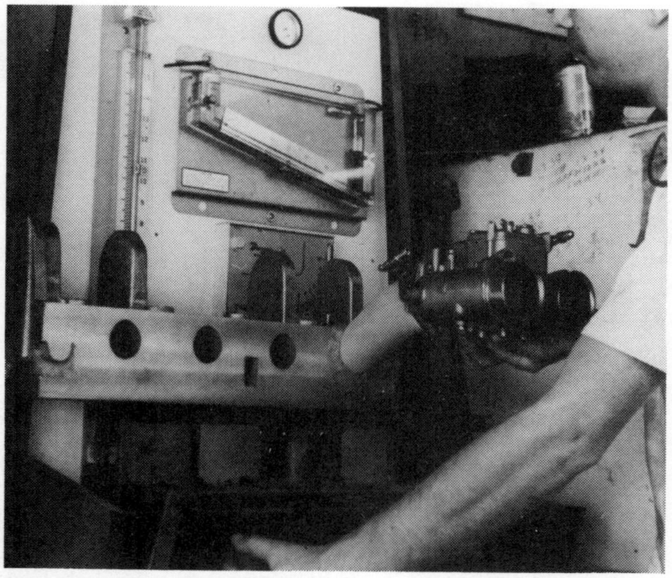

Flow bench and dyno work play an important role in any engine development. See the text for the flow bench comparisons between the stock and the aluminum head. Here, Dan Esslinger is testing a side draft manifold design.

heads they use Ford Motorsport's TRW valves and for their aluminum heads, the Del West titanium valves. Both manufacturer's valves measure 1.875 inches on the intake, and 1.600 inches on the exhaust. The reason for the titanium valves in the aluminum head is that the Esslinger head increases the RPM peak by some 500 to 700 RPM. The reciprocating weight of the steel valves becomes too great after 7,500 to 7,800 RPM.

Dan Esslinger cuts the valve seats at three different angles: 30, 45 and 60 degrees. The valves are ground at 45 degrees and chamfered at 30 degrees to bring the seat to the desired dimension.

Once the valves are ground, they are hand seated to provide a perfectly tight seal. Some solvent is added in the combustion chamber and compressed air is blown in the ports. If no air bubbles appear, the valve work passed the test.

The valve lengths in the Ford 2.3 are very critical in maintaining the proper valve train geometry and camshaft timing. If the valve is too long, the rocker will make contact with the valve near the inside portion of its contact pad, reducing the amount of lift available. If the valve is too short, the rocker will contact the valve with its tip.

At an extreme and with enough lift, the rocker can actually slip off the end of the valve. In addition, when the length of the valve is increased, the height of the valve end of the rocker also changes, introducing the rocker to the cam lobe earlier on the intake side and later on the exhaust. Incorrect valve length will also cause excessive valve stem and guide wear. Valve length and cam timing should be checked after a valve job is performed.

The valve keepers used on titanium valves should be ground so as not to touch each other when installed on the valve. The reason for this is to stop the valve from turning in the guide and the keepers, because titanium does not wear well in a rubbing situation.

Esslinger uses TRW valve seals from Ford Motorsports.

Valve Springs and Retainers

Esslinger uses Crane valve springs, retainers and camshafts exclusively. The Crane part number is 99884-8 for the springs and the retainers' number is 99967-8. Be certain to use the steel washers under the valve springs on the aluminum head, otherwise the springs will wear the seats on the head.

The Camshaft

We used a Crane cam FOR-300-8 on our motor. This cam gave a strong, wide power band which was our goal for short track racing. For long tracks, 1/2-mile or more or for road racing, Esslinger recommends Crane's FOR-320-10. The cam specifications are as follows:

	Lift	Duration	Lobe centerline
FOR-300-8	.510	300 deg.	108 deg.
FOR-320-10	.560	320 deg.	110 deg.

Valve lash is set at 0.010-inch and it is changed depending on conditions, needs and form of racing.

Camshaft Installation

Before installing the camshaft, apply a liberal amount of white grease to the bearings. Use caution when installing the cam not to damage the bearing surfaces.

Once the camshaft is in place, the sprocket can be installed. Esslinger makes an adjustable sprocket that allows timing the cam with pinpoint accuracy. These sprockets come in handy after a head has been milled, to reset the cam timing that was lost in the slack the belt had to take up.

Always use some gasket sealer on the camshaft pulley bolt. The camshaft is hollow, to allow the oil to travel through to the lobes. The oil passage is drilled from the front of the cam, and the pulley bolt plugs up the passage.

The camshaft on the 2.3 Ford can be timed to within 1/2-degree using a water level. First, level off the engine by placing the level on the back of the head and rotating the whole engine on the engine stand until it is level. Turn the camshaft until the number one cylinder lobes are in the up position, about 10 and 2 o'clock. Rotate the engine clockwise (normal engine rotation) until TDC is reached.

Use caution not to turn the engine to the left. If you do, the valves will hit the pistons (the pistons should have been left half way through their stroke when they were installed). Balance the water level on top of the two lobes.

Before setting the camshaft timing, set the engine level. Do not mount the side draft carburetors solid like in this picture. These carburetors were mounted this way to check for frame clearance in the car.

With the crankshaft at top dead center, level the number one cylinder lobes and align the timing marks on the pulleys. The Esslinger adjustable cam timing sprocket allow quick cam timing changes. Another advantage of the adjustable sprocket comes when the head or block has been milled, which changes the distance from the crankshaft to the camshaft and retards the cam timing.

After installing the camshaft, rotate it to make sure it spins freely. Then install the camshaft locking plate and check that the cam is still turning freely. Use plenty of high temperature white bearing grease on the camshaft bearing surfaces.

Smear plenty of camshaft assembly lube to the lobes. Arrow points to one of the valve adjuster posts. Note the large size of the oiling holes in the camshaft lobes. That's where much of the oil volume goes on an overhead cam motor.

Chances are that your water level is not wide enough to straddle both lobes, so use a small, flat plate on top of the lobes to set the level on. Gently rotate the camshaft until the lobes are level and install the timing belt tensioner.

To install the timing belt tensioner, first install the large bolt and push the tensioner in place, ready to accept the smaller adjusting bolt. At this stage it is easier to tighten the larger bolt to hold the tensioner in place, and then install the smaller bolt and the belt. The tensioner can be released once the belt is in place. Release the tensioner now, or the camshaft timing will change if the tensioner is released after the sprocket bolts are tightened. Double check your adjustment by rotating the crankshaft clockwise two turns to bring the camshaft back to its number one, up position. If you have a brand new sprocket, stamp a small punch line at the 0 mark on the sprocket—this will be your 0 degree mark for the camshaft. If the camshaft is being re-timed, set the mark on 0, or where you found the motor to work at its best.

Everything must be spotlessly clean before reassembly. Here Dan Esslinger is cleaning the valve guides with a rifle bore brush.

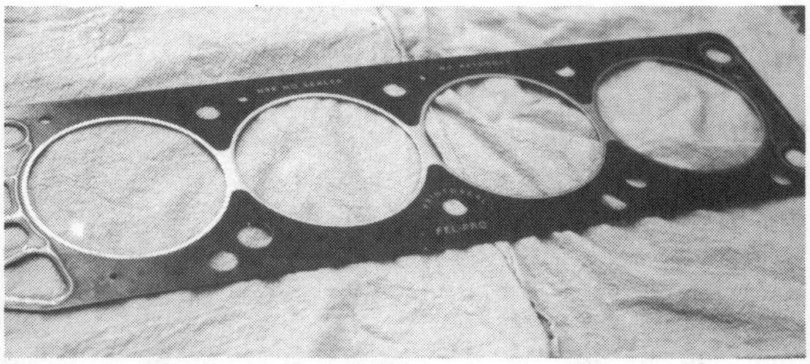

Esslinger uses Fel Pro and Ford Motorsport gaskets. Both gaskets provide good support around the critical cylinder edge area.

The Esslinger crankshaft pulley is graduated for setting the timing and its smaller size slows down the water pump. The smaller pulley size helps placing the engine low in the chassis.

The Camshaft Adjustable Sprocket

When the head is shaved, the camshaft drops closer to the crankshaft, causing the belt to loosen and the timing to retard. To correct the situation, Esslinger manufactures an adjustable camshaft timing sprocket. The sprocket will take up the slack in the belt, and it provides a quick method of setting the cam timing at the track. For more bottom end torque, the sprocket can be advanced up to 8 degrees, and the top end can be improved by retarding the sprocket by another 8 degrees. Maximum recommended movement is 6 degrees, either way.

The Exhaust System

For short track racing Esslinger and Cyclone recommend 1 3/4 inch primary pipes, 28 inches long, into a 3-inch collector. The outlet is 3 inches in diameter and 12 inches long. Remember that header tubing size is quoted in outside diameter measurements. Our primaries are 1.75 inches, minus 0.049 (wall thickness) times 2 equals a true inside diameter of 1.65 inches. This configuration works well for tracks up to a half mile in length. Longer tracks and road racing motors should use headers a couple of inches shorter.

Our Mini Stock headers were custom built by Cyclone and they are being used to jig-up for the production models. The collector rotates, which makes it easy to aim the exhaust in any direction.

We used Cyclone headers (see text for specifications). Building headers always starts with the farthest cylinder from the collector, in our case the number one. The tubes are joined in this jig so the finished headers will fit in the four-into-one collector

The Muffler

A good racing muffler will not hurt performance. We used a Supertrapp system with eight plates and we passed the noise control standards at our track. The power output was not affected. Supertrapp engineers tell us that a good muffler will not hurt power and under some conditions a good muffler can actually help power output.

The number 1 cylinder is farthest from the collector so it is joined to the closest opening in the collector. The number four cylinder, being closer to the collector, is joined with the farthest opening.

All the tubes were acetylene welded without any welding rod, and they were cut on a band saw. Smooth bends like these in the header pipes have no effect on power. The outlet pipe was mounted to pivot so it can be aimed in any direction.

Maintenance Schedule

Esslinger Engineering recommends the following maintenance and inspection schedule for the Ford 2,300 OHC.

1) After each race: Change oil and inspect for burnt color and odor. Check the level before draining.
2) Inspect and clean filter. Replace if you use a non cleanable filter.
3) Inspect and replace spark plugs.
4) Check valve lash and compare against last setting.
5) Inspect camshaft for wear.
6) Inspect timing belt and water pump belts. If you run a dry sump system, inspect the pump drive belt.
7) Check for coolant, oil and fuel leaks.
8) Check throttle linkage for full range, return and possible binding.

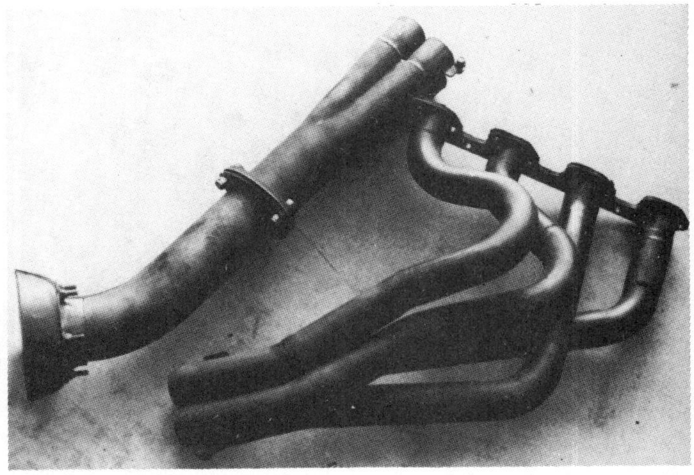

The finished headers are light and strong. The exhaust noise is muffled by a Supertrapp muffler.

9) Check intake and exhaust manifold mounting bolts.
10) Clean air filter (s) if necessary. This is a must if you run in the dirt.
11) Check ignition timing.
12) Check oil pan and valve cover bolt tension.
13) On the last lap of the last race of the night, take a good look at the oil pressure and temperature. Compare these figures with the prior race numbers. Do a cylinder leak-down test, note and compare figures against last test.
14) Check spark plug wires for looseness, cracking or signs of cross firing.
15) Check all wiring.
16) Half season: Check rod and main bearings.

Torque Specifications and Clearances

Torque

Cylinder head bolts	90
Main bearing cap bolts	100
Connecting rod bolts	65
Flywheel bolts, ARP Pro	112 to 115 with oil
ARP High Performance	78-80 with oil
Header bolts	25
Intake manifold bolts	30

Clearances

Piston to bore	.006
Piston ring end gap	.012
Main bearings	.003*
Connecting rod bearings	.003*
Crankshaft end play	.004 to .007
Connecting rod side clearance	.012 to .015
Wrist pin	.0008

*Use .0035 if the motor is to be broken in on the track.

Chapter 14

The Fuel System and Carburetion

A real safety fuel cell is a necessity in a racing car. A passenger car fuel tank in a race car is a time bomb just waiting to go off. A real fuel cell is designed to sustain blows and punctures without tearing, many times the force that a passenger car tank can survive. In addition, if a fuel cell does puncture, it has an inner liner that is designed to hold the fuel and inside the bladder is foam that will slow the leakage. Fuel cells are equipped with upside-down valves in the breather and flapper valves in the filler neck.

We referred earlier to a real fuel cell. ATL Competition Equipment refers to a fuel cell as: "A flexible, rubberized, deformable bladder, housed in a metal container. The real fuel cell also contains some baffling, such as in the form of foam and it contains a hinged flapper in the filler neck and an up-side-down valve in the breather". Based on this definition many racing cells used in stock cars are not real fuel cells.

The true culprit is really the rule book wording by the sanctioning bodies. "Approved" has become a meaningless standard adopted by most short track organizations. Rely only on cells carrying the approval of the larger sanctioning bodies—NASCAR, SCCA, FIA, USAC and the drag racing and off-road bodies—and buy the cell from reputable companies such as ATL or FBI.

When purchasing a cell, always ask to see test results and beware of "good deals". The three parameters to look at are: tensile strength, tear resistance, and puncture strength. Furthermore purchase a cell with a fill plate made of sheet steel or aluminum, rather than the cast material that can crack in a crash.

Fuel Cell Installation and Design

For short track application an eight gallon tank should suffice, while for road racing, you might want to go to a twelve gallon cell and larger yet if you plan on running the longer professional races.

The cell should be mounted between the frame rails and away from the rear bumper and rear end. The cell should be centered within the frame so as not to influence handling as the fuel load drops. In addition, the fuel cell should be protected at the rear by a hoop, to keep cars from slipping under the rear bumper and into the tank. Many sanctioning bodies have certain rules regarding the mounting of the cell. As a rule, the following guidelines should do:

Build the tank frame out of 1 1/4-inch steel angle, by 3/32 inch (the type of angle iron bent on a break is of better quality and will weld better). This material can be recognized by the rounded shape of the angle, compared to the sharp corner of the regular material). Run two 1-inch by 1/8-inch steel straps from front to rear and two more straps from side to side, under the cell.

In addition, some organizations require at least two more straps across the top of the cell, which is a good idea.

Good installation of the fuel cell, the pump and the filter. Be sure the rear end clears the pump when it goes into bump. The breather hose on the tank is equipped with a one-way valve to prevent fuel spillage should the car roll over. Note rear end breather and filter at the right of the photo.

Mount the top straps so they can be removed to allow for servicing the tank. The steel box should be made of at least 20 gauge steel with a one inch lip all the way around it. Some builders weld a box in the car and then they install another steel enclosure that is bolted in place, in the first container. You can never be too safe!

Run a breather vent with the one way valve sold by ATL. Place the valve vertically in the line and extend the hose to the very back of the car and away from the brakes and headers (rear engine cars could have this problem with the headers).

In our car we used an ATL Saver Cell, part number SA112. The cell holds 12 gallons and we also used the ATL container, fill plate and breather upside-down valve.

If you run a full bodied car, with the rear bumper area enclosed as required in road racing, get the straight filler neck with extension and mount the filler cap on the outside of the car. For a Mini Stock, get the 45 degree fill valve , ATL part number TF-184. The angle of the filler neck will facilitate filling the fuel from the rear of the car.

ATL SAFETY FUEL CELL Principle of Operation

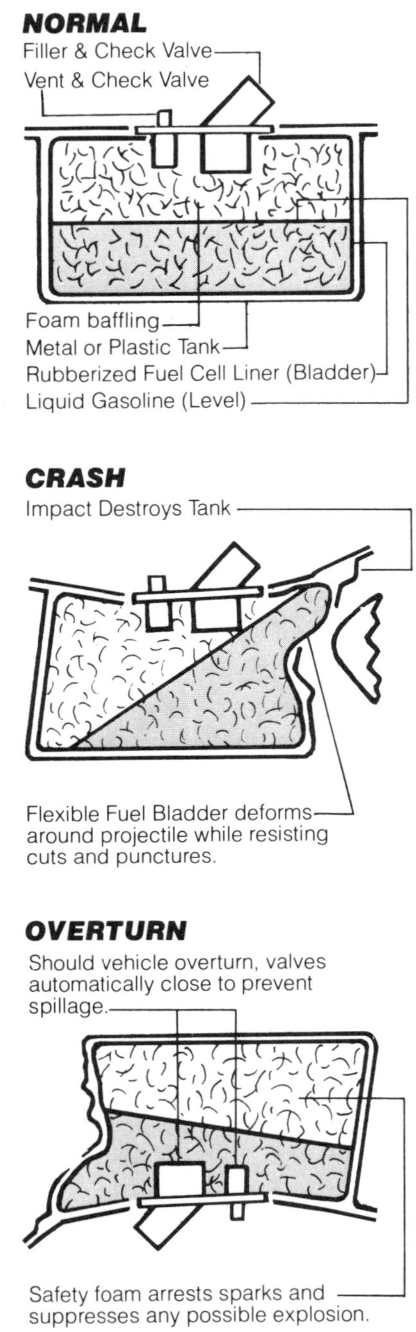

Plumbing

Redline, the Weber distributor in the US, recommends dash 8 lines (1/2-inch) at least to the regulator and dash 8 or 6 from the regulator to the carburetors.

The fuel system starts at the fuel cell. We should mount the fuel pump close to the fuel source for maximum fuel supply. We used the Ford Motorsports heavy duty fuel pump, part number M-9350-B111. It pumps 75 gallons per hour at 7 PSI.

The fuel filter is an Oberg motorcycle oil filter. These filters have a large filtering area and they are easy to inspect. Of course the fuel pressure regulator should be installed as close to the carburetors as possible. The Holley regulator sold by Ford Motorsport, part number M-9968-B111, splits into two outlets. This allows the use of the regulator as a T connector and to go into the carburetors from the regulator.

Run a bulkhead fitting to go through the rear fire wall. Never run fuel lines directly through sheet metal without using a bulkhead fitting. The consequences of a fuel leak are too great to take the chance. Earl's and Baker Precision carry a complete line of braided hose and fittings for this purpose.

From the bulkhead fitting the line runs through the driver's compartment. At first look this might seem dangerous, but the driver's compartment is the best protected area in the whole car and braided line is safe enough to use around the driver. Nevertheless, do not install the line in between two parts that could pinch the line in case of a wreck (such as between the seat hoop and the frame, or on the outside of the frame).

We mounted the fuel pressure regulator on the fire wall, near the steering shaft. The dash 8 fuel line goes in vertically under the regulator off the 90 degree bulkhead fitting. Each carburetor is fed individually from the regulator. To keep the fuel from picking up heat from the engine compartment, wrap the fuel lines with Earl's Flame Guard sheath.

Some carburetors are equipped with two fuel inlet tubes, to feed itself with fuel and then a second carburetor. Try to use, instead, a set of carburetors with the fuel inlets facing each other so each carburetor can be fed individually with fuel. Going in one carburetor with fuel and into another with the remainder of the fuel could cause some temporary fuel starvation in the second carburetor.

Carburetors

We chose Webers for our car based on Esslinger's recommendation. We used the 48 IDF downdraft, because of distributor and frame clearance problems. However, if frame and distributor clearance had allowed we would have used a sidedraft model. The straight sidedraft manifold flows much better than the downdraft model and it is worth five to eight horses on the dyno, but in order to use this unit, an angled distributor, or crank triggered ignition, must be used. A sidedraft manifold, made by Cannon, part number 0046, is available from BAP/GEON stores which will fit in a car with narrow clearance and it clears the distributor. In order to clear the distributor, the manifold is bent in a tight S shape that restricts the flow. So, with the steel head we are limited to the low flowing sidedraft manifolds or the downdrafts. With the steel head we are limited to about 7,500 RPM for peak power

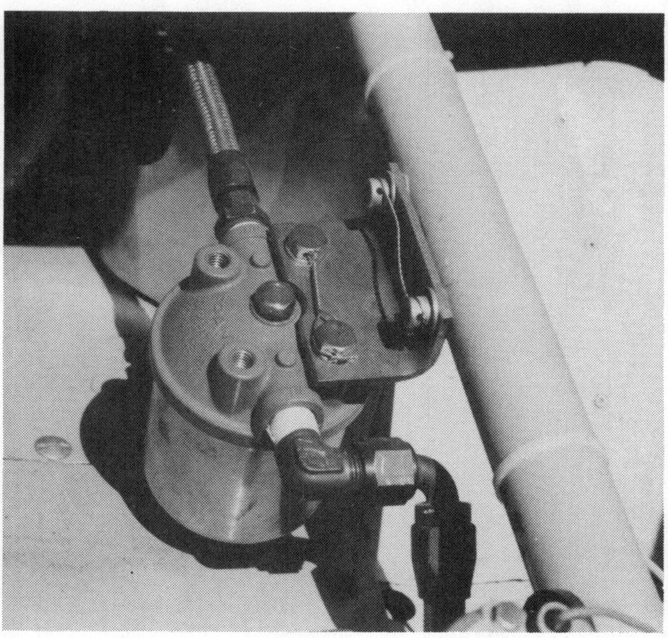

This fuel filter mounting shows a lot of thought. The lines are braided with stainless steel, the pipe fittings were wrapped with Teflon tape to prevent leaks and the fasteners were safety wired.

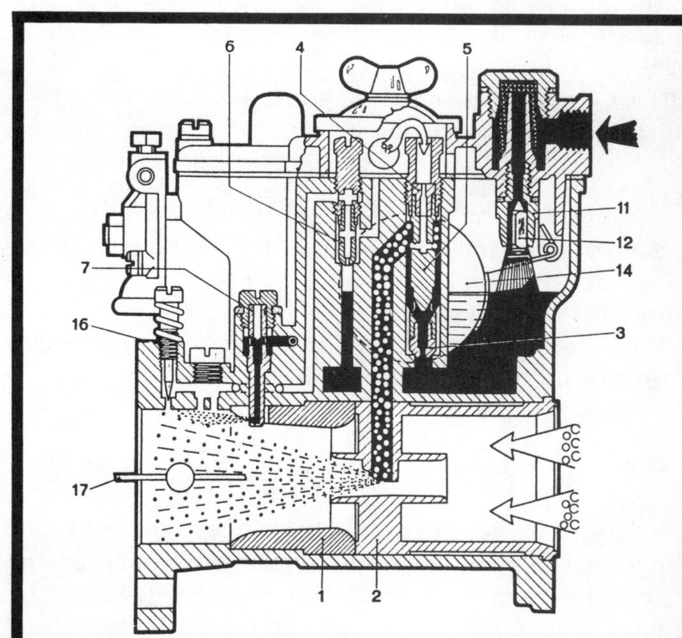

Section through a Weber DCOE series carburetor: 1) main venturi, 2) auxiliary venturi, 3) main fuel jet, 4) main air jet, 5) emulsion tube, 6) idle speed fuel jet, 7) accelerating pump jet, 11) needle valve, 12) valve needle, 14) float, 16) idle mixture adjusting screw, 17) throttle valve.

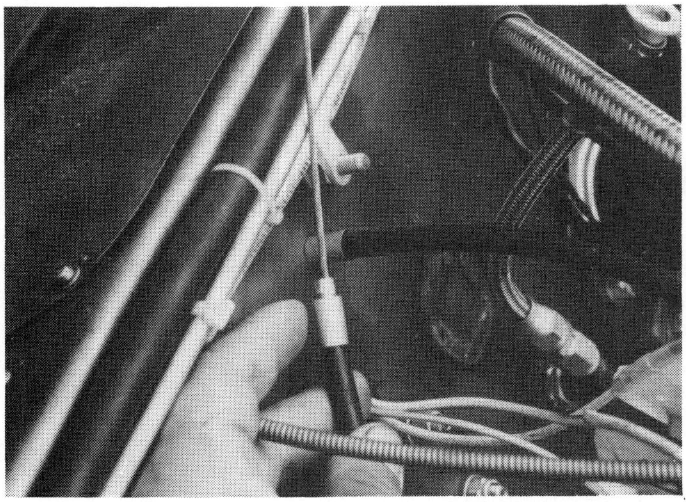

The end of the throttle cable can be threaded and run through the firewall as shown here. Use a low profile lock nut to secure the cable end to the firewall.

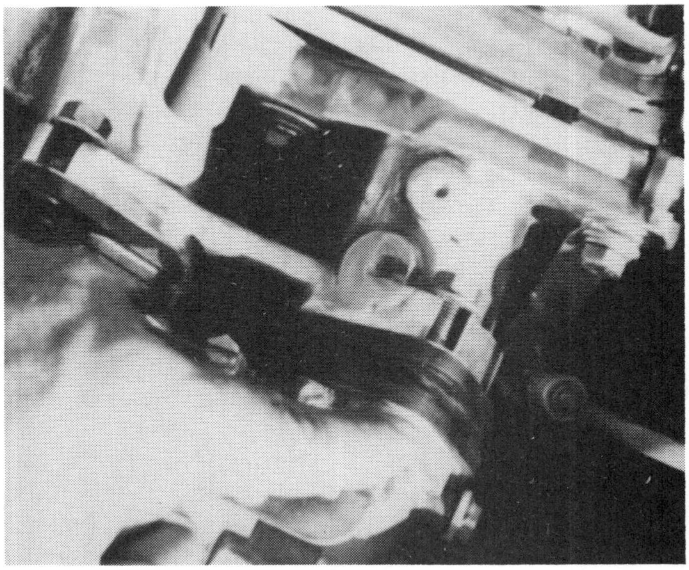

The rubber insulators mentioned in the text.

and Weber literature dictates a carburetor of 45 to 48 for maximum power at this RPM range. Esslinger has shown on the dyno that with the steel head, the 48 IDF carburetors, with the downdraft manifold, are ideal in the power range needed for short track and road racing. With the Esslinger/Motorsport aluminum head, the picture is somewhat different. With the greater breathing ability of the custom head and the custom manifold designed by Esslinger, it is important to supply the engine with all the carburation the engine can use. All efforts should be made to use the sidedraft manifolds, with 50 DCO Webers. The aluminum head, sidedraft manifold and the 50 DCO carburetors combination is worth over 40 horsepower over the steel head with the downdraft carburetors.

Carburetor Jetting

With our motor we ended up with the following jetting in the 48 IDF's: Venturis, 39mm, main jets 155, air correction 180, emulsion tubes F2.

With the aluminum head and the 50 DCO's you can start with the following: Venturis 46mm, main jets 160, air correction jets 160, emulsion tubes F7.

Carburetor Mounting

Sidedraft carburetors must be rubber mounted. The length of the manifolds and the leverage applied by the carburetors amplifies the vibrations produced by the engine. The vibrations transmit to the jets and the fuel inside them. The fuel is churned in the jets and fuel flow is disturbed, causing mixture variations. BAP/GEON stores sell rubber mounting kits for the sidedraft Webers. The most effective method of dampening vibrations is to install the intake manifold rubber dampeners obtainable

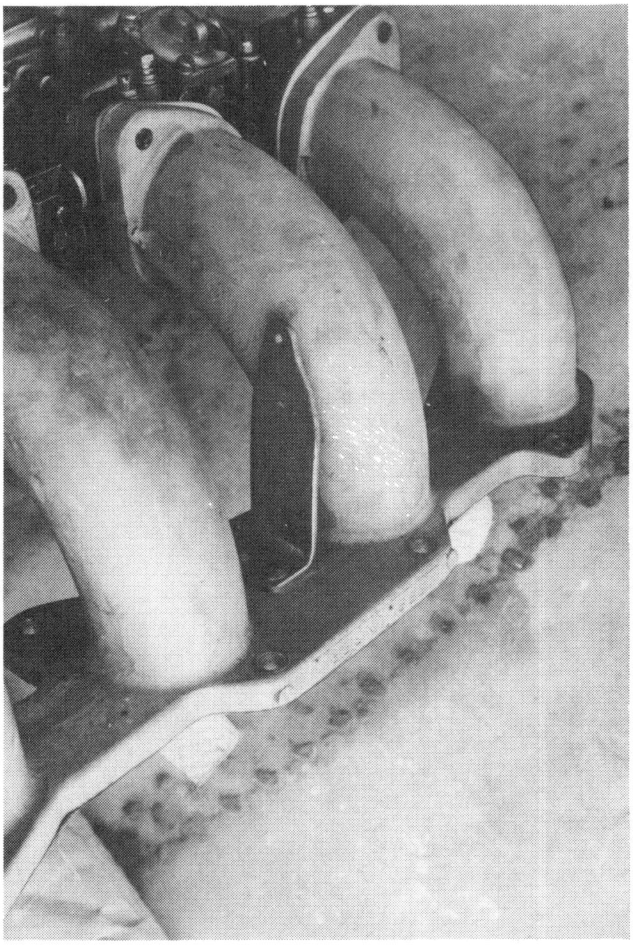

The Esslinger downdraft manifold and throttle cable bracket. The bracket can be made out of 1/8-inch steel plate. Be sure the bolts used to mount this part end flush or below the manifold face or they will bottom on the head. Use gasket sealer on the bolts to prevent leaks.

at Nissan dealers. The rubber spacer part number is: Nissan Motorsports 16174-20100 and the gaskets' part number is 16176-14610. These spacers will fit 45 and 48 mm Webers and for use with the 50 DCOs the bore of the spacer will have to be enlarged slightly.

The downdraft carburetors are not as influenced by vibrations, but rubber mounting would still help performance because of the heat barrier the rubber would provide for the carburetors. A cool incoming charge will produce more power than a hot one. Keeping the carburetors, fuel and the incoming air as cool as possible should be a priority. Building a cold air box around the manifold, carburetors and the air filters is highly recommended. The carburetors and the air filters must be enclosed in the same box, so as to receive the same air supply. If the air pressure at the intake is different from the pressure in the float bowl, the air-fuel mixture will be drastically disturbed.

Fuel Pressure

Recommended fuel pressure for Weber carburetors is from 2.5 PSI to 4 PSI. Try to keep the pressure as low as possible without causing any fuel starvation. 3 PSI is a good starting point. The pressure should be checked with the engine running.

Intake Air Horns

The purpose of air horns on a racing engine is to prevent turbulence at the air inlet, which would cause varying mixtures at different speeds and uneven cylinder charging. The air horns are also used to reduce the effect of the blow back through the carburetor. Air horns build up harmonic waves within their throat and the carburetor. These waves have an effect on the overall power out put of the motor and on the power band. The length of the intake horns acts in the same fashion as the intake manifold length does on the location of the power on its range. A longer air horn will show some power increase at the lower part of the power band, while a shorter horn will show an improvement at the higher range of power. An increase in performance at one end of the power band will be at the expense of the other end of the spectrum.

The air horn opening needs a minimum of 1 1/2 inches of clearance between its end and the air filter to perform properly.

Air Filters

A good air filter can actually improve horsepower. The days of removing the air filter to improve the power are gone. K&N Filters has shown that controlling the charge pulses at the air horns improves intake charging. The air passing by the air horns disturbs the incoming flow and it can lower power output. A good air filter will also protect the engine from dirt and debris. Rubber coming off tires, fiberglass from wrecked cars and track dust will tear up a motor in a hurry, not to mention that a chunk of rubber can cause the throttle or an intake valve to stick open.

Choose a washable filter such as the K&N elements. These filters can be reused and they have proven themselves superior in horsepower tests over the paper elements.

Gasoline

Choose your gasoline carefully for the Ford 2.3. Esslinger has tested all the commercially available racing fuels and their dyno figures have shown that fuel quality is the limiting factor to compression ratio. Since compression ratio is directly related to power, we can say that you can only go as fast as your gas will let you. Esslinger uses only So Cal Blue gas. The Blue gas is 125 octane and it is the only "pump gas" that allows the 2.3 to run 13.5 to 1 compression ratio and still show some power increase over the lower compression ratio. Good gasoline also improves reliability by reducing, or eliminating, detonation and preignition.

Chapter 15

The Dry Sump and Plumbing

The Dry Sump System

A dry sump system is desirable in a race car for two important reasons: 1)reliability and, 2) performance. A dry sump allows you to run larger quantities of oil, contained in a separate reservoir, since the oil is carried in a separate tank and not in the pan, and the oil pan can be designed very shallow, allowing the engine to be dropped down low in the frame.

A dry sump pump draws its oil from the remote reservoir, giving it a constant supply, without any pressure losses in turns or during braking or acceleration. The oil pan designed for a dry sump system is equipped with the necessary baffles, scrapers and screens to scavenge all the oil out from around the crankshaft. This prevents the oil from being churned by the crankshaft and the rods. The oil is not as aerated, improving pressure, cooling and volume. The churning process in a wet sump pan also causes great power losses and it throws the crankshaft out of balance. The sump tank de-aerates the oil before it is sent back to the pump.

Basic Design

The whole process starts in the engine, after the oil has done its job of lubricating and it finds its way down from the head and the crankshaft into the pan. Three pick- ups are used to "scavenge" the oil, one from the valve cover area and two places in the pan. In our motor, the oil return passages in the head have been plugged and re- drilled to 0.050-inch to allow oil drain back after the motor is shut off.

If these hole are not plugged, under acceleration the oil would drain back directly on the rear crankshaft flange,

Fitting at the rear scavenges the oil from the cam cover area. Fitting at the front is for the breather line that comes from the sump tank. The stock breather cap is used as the breather outlet for the valve cover. Another vent is used from the block, because the oil return holes in the head were plugged off to prevent the oil from dripping on the crankshaft, making the crankcase air tight. Small holes were left in the oil return holes to allow oil drainage after the engine is shut off.

The Patterson oil tank. The oil drops in the tank, swirling on top of the umbrella shaped baffle and then through the holes. This dearates the oil. The oil finds its way down the tank and it enters the large vertical pipe through the holes which remove the last of the air bubbles. The oil is scavenged by the pipe in the tank half on the right. Just below the scavenge pipe is the connection for the tank heater.

Modine recommends running the oil pressure line in the bottom of the cooler and out the top. This way all the air will bleed out and the oil will not drain out of the cooler into the engine when it is shut off.

robbing several horses right when they are needed the most.

At this point it is imperative to point out one of the peculiarities of the Ford 2.3. This is actually a characteristic of overhead camshaft motors in general. The camshaft is drilled through each of its lobes to allow liberal amounts of oil to flow to lubricate the critical camshaft lobe-rocker area, actually the rocker arm needs the most oil because it is in contact with the camshaft lobe a great majority of the camshaft rotation, while the lobe only contacts the pad once per revolution.

These holes in the lobes squirt oil constantly without any restriction while the connecting rod and main bearings, for example, run on very small clearances (0.003 inch) that restrict the flow of oil. This point is very important for two reasons: 1)The 2.3 motor pumps every large amounts of oil and most of this volume ends up in the valve cover. If not scavenged properly, this oil will find its way down on the crankshaft, or worse yet, out the breather. The 2.3 Ford pumps more oil than the small block V8 motors do! 2)The pump must be chosen according to this fact. A pump used on a small push-rod motor just will not do on an overhead cam motor. Along the same lines, the plumbing must be designed to carry large amounts of oil and a filter with a large filtering area must be used to prevent any flow restriction.

Once the oil has escaped into the pan and the valve cover, the suckers go to work, aspirating the oil into the pump. There the oil is pressurized and channeled into a manifold. The oil is then taken from this manifold to the sump tank, via a large oil line (minimum of 3/4 inch). The oil is dumped in the sump tank, or more precisely, on the baffles in the tank. There the oil swirls around on the baffles to de-aerate itself.

Once the oil reaches the bottom of the tank, it is sucked back to the pump. At the pump the oil is again compressed and sent to the filter, the cooler and then to the engine.

The sequence described should be followed religiously.

Plumbing

Plumbing the dry sump system is just as critical as the choice of the components. Peterson Automotive Research, the pump supplier for Esslinger, recommends the following: 1)Always run screen filters in the scavenge lines. This will prevent large pieces from entering the pump, especially if the engine blows. 2)Never use braided lines on the suction side of the pump, both for the scavenge and the return from the tank to the pump. Use Aeroquip's 2556 black suction hose, or FC 300 blue hose. Both these hoses are available from Baker Precision. These hoses are designed specifically for low vacuum and high temperature.

The braided lines can take a lot of pressure, but very little vacuum before they collapse. Peterson reports that the braided lines are most prone to collapsing when they are heated. Since oil can come out of the engine at anywhere from 200 to 300 degrees it is easy to see that this type of hose just does not suit our application.

3) Never run the filter or cooler in the line from the pump to the tank, or in the suction line from the tank to the pump. The line from the pump to the tank must have an absolutely unrestricted path, because this line carries not only oil but also a large amount of air which can as much as double the volume of the oil. The cooler does not belong in that line for a second reason—aerated oil does not cool well.

The return line to the tank should be a minimum of 3/4 inch (dash 12).

The suction line from the tank to the pump should also

be unrestricted and of -12 minimum. In addition this line should be able to sustain vacuum drops. The suction line should not be of the braided type because braided hose does not hold up under vacuum and heat, and it may collapse. Even if the hose does not completely collapse, a small restriction is enough to ruin an engine. The oil filter absolutely must not be placed in this line. Pumps are designed to push fluids, not suck them. Any restriction in the suction line, such as the filter, would become a tremendous obstacle for the pump to overcome. Believe me on this one—I already ruined one engine because I ran the filter in-line from the tank to the pump.

Peterson recommends the vacuum hose for all three of the suctions out of the motor. The first hose comes out of the valve cover, into a Moroso in-line screen and into the pump. The other two suckers come out of the Canton pan. Canton installed the screens in the pan which makes things difficult to clean, but then again we would have been hard pressed to install in line screens because of the lack of space available between the pump and the block.

To join the three scavenge lines into one at the pump, we had a stainless steel manifold made by Leon Ruther Enterprises. Leon specializes in stainless steel fabrication and he makes these manifolds for several Grand National teams. Take all the dimensions in consideration before ordering a manifold for the pump. Keep in mind that if the scavenge line going back to the tank is going to go back directly, it will have to pass over the pressure-out fitting. Try to have the manifold made so it will allow you to remove the pressure-out fitting, without having to remove the manifold. This manifold can be installed in a remote area and it can be joined to the pump by braided lines. This would mean the extra expense of the line, six hose ends and three AN connectors.

The top line is for the tank breather. It is a -12 line and it ends at the front of the valve cover. Line below that is the line for the oil scavenged out of the motor and this line is also a -12 (3/4 inch). This line should be large because the oil scavenged out of the motor is not only very hot (which causes expansion) but is also loaded with air, causing its volume to increase dramatically. The bottom line carries the oil from the tank to the pump. It should be a -12 line and since it operates under negative pressure it must be made of the black suction hose material. If the braided line material is used, it may collapse when it gets heated. Also, since pumps are designed to push fluids, not to pull, never run the oil filter in the tank suction line. The restriction caused by the filter will cause a drastic pressure drop. Note how the tank is mounted.

Run the filter between the pump and the cooler with 5/8- inch (dash 10) braided hose. The filter should be in-line before the cooler to clean the oil before it enters the cooler and if the engine should blow, the cooler will be protected from debris.

The oil cooler should be plumbed in a cross flow fashion if possible. Our Modine cooler, part number 1B-111 has four fittings, two on the bottom and two on top. Go in either on top or bottom (ideal method is in the bottom and out the top, but Modine found very little difference in going in the top and out the bottom) but come out at the diagonally opposed corner. This will ensure the oil has taken the longest path through the cooler, getting maximum cooling. This method only applies to coolers with tanks, not the tube type coolers.

The breathers are extremely important on a dry sump motor. Without good breathing, the pump would create a vacuum in the crankcase and a drastic power drop would occur (and probably a quick pump failure). The breather can also be used to cool the under-the-valve-cover area and especially the valve springs. Peterson notes that the valve springs run very hot because of the repeated work they do. In addition, the exhaust valves are constantly vented with red hot gases and this heat is dissipated, in part, through the valve stems and eventually, the springs.

The pump is constantly running, whether oil is present or not, causing it to pick-up large amounts of air to be added to the air already contained in the oil. This air is carried in the oil and released in the tank. If a line is run from the tank to the front of the valve cover and another line comes out of the back of the rocker cover, we have created a small air conditioning system for the valve springs. Once the blow-by exits the rocker cover it must be routed to a catch can, which in turn must be vented.

For a catch can we used a Moroso power steering fluid container. The line from the rocker cover goes in the bottom and the breather is attached near the top. For a breather we used a K&N breather, with a K&N washable filter. Remember to use a good filter at the breather, because the air that enters the system will end up in the motor. The breather line comes off the rocker cover at the stock filler cap, although if you run the aluminum Motorsport rocker cover, it would be possible to weld a 90- degree fitting on the top of the cover and attach the breather line via an AN fitting.

The scavenge out of the rocker cover area can be tapped in the head or in the cover itself. Insert a 90-degree bulkhead fitting through the cover and aim the long end down toward the head. Some trimming might need to be done, but this set-up has worked well. A hole can also be drilled and tapped in the right rear of the head, just below the valve cover. By installing a 90 or 45-degree fitting, pipe

The oil pressure adjustment on the Peterson dry sump pump. Just a half turn will show a change in pressure.

After the pulley on the sump pump has been adjusted, it is imperative that the three Allen screws be tightened. Peterson recommends using some Loctite on the Allen screws.

thread at one end and AN at the other, you can run a dash 10 line and a Moroso in-line screen down to the pump.

At first look, it might seem logical to install a suction line at the front of the rocker cover and one at the rear. Peterson says never to tee-off any suction lines on a dry sump system. The system will follow the path of least resistance and it will suck air before sucking oil.

It is hardly possible to plumb a system without any 45 or 90-degree fittings but keep these to a minimum. Avoid at all cost the cast angled fittings. These fittings severely restrict the flow of oil. Instead of running a 90 in a casting, use a straight pipe fitting in and a 90- degree hose end.

The Pump

Peterson Automotive Research supplied the pump for our motor. The pump (part number P4105) is one of the largest made by Peterson and even though it was thought to be an over-kill in the beginning, it turned out to be just the right size for our oil hungry 2.3. Peterson pumps come in three stage (two scavenge lines) and four stage (three scavenge lines). A V8 does not have the amount of oil we have in the 2.3 Ford valve cover so the V8 guys can get by with a three stage pump. With the amount of oil we have in the valve cover, we must scavenge it out efficiently with the third scavenger.

Peterson's pumps come in either right or left side mount. We mounted our pump on the right side of the motor. The pump would be more conveniently located on the left, but we could not clear the steering shaft and the frame lower crossmember. If your motor has a different setback, or your steering rack is in a different location and you can fit the pump on the left, do it. It would be nice to be able to mount the pump on the left and mount the tank on the fire wall, to minimize the length of the hoses.

The Sump Tank

We used the 3 gallon dry sump oil tank from Patterson Enterprises. If you have a choice you might as well go to a four or even five gallon tank. The extra capacity can only help keep the oil cooler.

Get a tank with a fitting for the oil temperature sending unit and a heater.

The tank should be mounted so the top of the oil level is higher than the pump to allow some gravity priming. The tank should be mounted as close to the pump as possible to reduce oil drag in the lines. We tried to fit the tank behind the driver's seat but there just was not enough room. The most feasible location is in the back of the passenger seat. Another possible location is at the firewall, facing the left side of the car, behind the left front wheel. This location, however is very vulnerable in case of a crash. Mount the tank solidly. Five gallons of oil, the tank and its hardware will weigh around thirty pounds.

Before installing the tank, it should be taken apart and all the punched holes should be de-burred. Clean out evrything before assembly.

A 3-gallon tank should only be filled with a little over 2 gallons, after the system has been primed. The best way to do this is to place 2 to 2 1/2 gallons of oil in the tank and measure the level with a dipstick. Start the engine and bring the oil pressure up. Shut off the motor and check the level again. Add oil until the level is back up to the original mark. Note how much more oil was added and add that figure to the original figure for the total amount the system will take next time it is filled. Make a permanent mark on the dipstick.

It is important to pre-heat the oil via a sump tank heater, such as the ones sold by Patterson. Even in warm climates, racing oil flows very slowly when at room temperature as cold oil flow is very slugish. The pump has to work extra hard to suck the oil up to its rotors. The oil pressure will take much longer to come up when the motor is first fired up, causing possible damage. The tank heater eliminates all these problems.

The Filter

Esslinger Engineering uses the System 1 filter exclusively. This filter can be inspected and it has the added advantage of being a spin on which facilitates removal and installation. The cartridge style stainless steel screen element allows quick reading of the engine condition. Most importantly, the System 1 has almost three times more screen area than other screen type filters.

The System 1 filter has shown a dramatic improvement in oil pressure over many of the high performance filters currently on the market. Esslinger experienced a 25 pound improvement over another popular brand using the System 1. Remember that, considering the amount of oil the Ford OHC pumps. Any restriction will have a major effect on flow and pressure.

When assembling AN fittings and braided lines, use aluminum jaws in the vise and wrap some tape around the fittings to protect them from damage by the wrench.

Absolutely avoid cast 90 degree fittings like in this photo. Plan the line routing so it will use as many straight fittings as possible. If a 90-degree fitting is necessary, use the bent tube type. They do not restrict flow as much as the cast fittings because of their smooth radius tubes. In this photo, the battery and the oil filter are in a very vulnerable position in case of a hit.

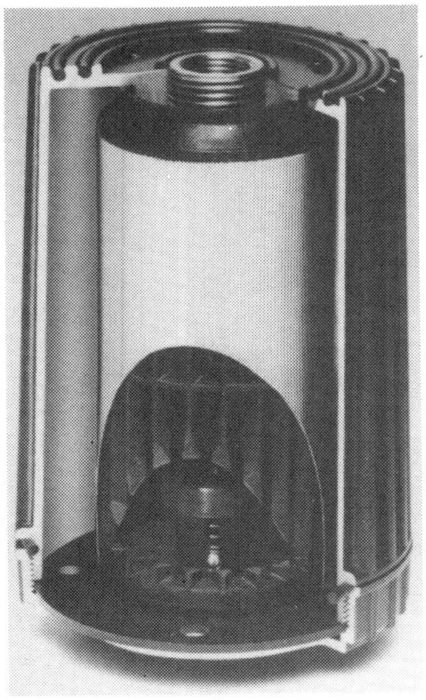

Esslinger uses the System 1 oil filter. The Ford overhead cam motors flow large amounts of oil and this filter has shown the least amount of pressure drop on the dyno. The System 1 has almost three times the filtration area of the other filters of this type.

The filter should be mounted between the pump and the cooler and again, try to minimize on the angled fittings.

Plumbing

Plumbing a whole car with braided lines and AN fittings can get expensive. However if you consider the safety factor, the reliability and longevity, the price doesn't seem so high any more.

All our plumbing needs came from Baker Precision and we used Earl's lines and fittings (except for the black suction line and its brass fittings which are made by Aeroquip). Earl's fittings are easy to assemble and they will not create any flappers in the hose.

Measure out your hoses and make a detailed drawing of the whole system before ordering the hoses and fittings. Better yet, if you can, take your car to Baker or Earl's and plumb it right there. This will probably save you many trips, shipments and phone calls.

To cut the braided lines, we found it best to use an abrasive cut-off saw. Wrap the line tightly with some duct tape and draw a line with a felt pen where you need to make the cut. Apply very light pressure on the saw or the braiding will not cut clean and you will have a hard time installing the fitting. Use some motor oil on the rubber before slipping the fitting on and always apply some anti-seize compound to the threads of the fittings.

When the assembly is complete, wash out the hoses with clean solvent, blow dry and plug the ends with the caps sold for this purpose.

Chapter 16

Electrical System and Ignition

The Electrical System

The wiring and the electrical system of a race car are often neglected by the builder. The wiring and electrical system are just as critical as any other component of the race car, and have an effect on safety, performance and reliability. All the system needs is one break in the chain and the car stops, or worse, burns. Performance-wise, a properly wired racing car will not have any electrical interference, voltage drop or ignition misfiring.

Wiring Basics

Electramotive, in El Segundo, Ca. is now the racing arm of Nissan, USA, but John Knepp, the owner, has an impressive background in race car wiring and electrical systems. Electramotive sells all the wire, connectors, terminals and tools needed for a professional wiring job.

Some of the basic rules to follow are pointed out by Electramotive: 1)Use only aircraft airframe wire. This wire only comes in white, so if the wire is shorted-out it will turn brown and the culprit can be easily spotted, but it will not burn through to cause a short. This wire has high abrasion and heat resistance. The airframe wire also has excellent insulation, so it can be tied in bundles without getting any cross-feed.

2) Use aircraft connectors with double crimp insulators. The terminals crimp the wire and the cable, so vibrations will be taken up by the whole cable, not just the wire. Also this will prevent the wire from tearing out of the terminal if it is accidentally tugged on. If you are going to use the auto store type of terminals, at least do not use push-on connectors, but use the eye type connector instead. The push-ons can slip off with vibrations or they can be accidentally knocked-off. Worse yet, the cheap push- ons will slip off in a wreck, causing sparks and risking a fire.

Forget about soldering cheap terminals. The solder will form a blunt end for the wire to go into and this will create

Electramotive sells high performance switches, wires, circuit breakers and connectors. Sensor and wire at the top of the photo shows method of epoxying connections to relieve the tension from the wire.

This type of connector is water resistant and vibration proof. Note the double crimped connections on this Electramotive car.

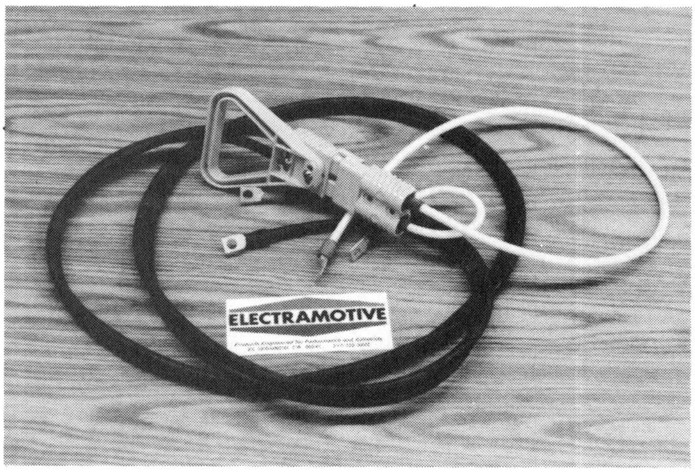

This switch is used to jump a race car battery. It can be disconnected quickly, and using this switch will prevent dead batteries.

a stress riser that may eventually cause the wire to break.

The aircraft double crimp terminals require a special crimper, available from Electramotive.

3) Switches should be water-proof and of aircraft quality. This is doubly important if the car is going to be raced in the dirt. Use switches with screw-on terminal posts, not the push-on.

4) Use a circuit breaker to shutoff the system if something shorts out.

5) Use continuous lengths of wires. Avoid splicing wires with butt connectors, and especially avoid making connections by twisting the wire and wrapping it with electrical tape.

6) When soldering, use a non-acidic core. Acidic solder will eventually cause corrosion in the wire.

Battery

Get the best battery you can find. A small truck or tractor battery such as the Sears 9601 or 9613 tractor type batteries is probably your best bet. Find the highest amperage battery that will fit in the location you have determined for your car.

Run the battery ground to the frame, not the battery hold-down clamps. The ground on the battery is for the whole car electrical system, so be sure it is of good design. This is especially important if you use an electronic ignition system. These systems need perfect grounds and a poor ground can create some erratic problems that are difficult to trace.

Try to enclose the battery in a marine application plastic box. These boxes are available from auto and marine parts stores. This will prevent acid leakage from getting on the car and the box will catch any acid that may have spilled in a wreck. It also affords extra protection to the battery.

We placed the electrical panel in this position so the driver could reach the switches and for ease of servicing. The master kill switch is easily reached by the driver and by people on the outside of the car.

Battery cables and terminals should also be of good quality. Use Electramotive's 4 gauge cable for the battery's positive and ground cables, or at least use some fine wire welding cable. Always use copper wire for the battery cables. Aluminum cable is lighter than copper cable but it does not have the conductivity that copper has and it may cause starting problems. The terminals should be crimped on with a special crimper. The crimping can be done by Electramotive or truck repair shops. Cover the terminals with heat shrink tubing to protect them from the elements and corrosion. Install the battery as close to the starter as possible and keep the cable as short as you can to prevent voltage drop.

It is a good idea to use a battery shut-off switch such as the one available from Moroso. The Moroso part number is 74100. These are required by IMSA and SCCA. The cut-off switches are installed within the driver's reach to facilitate complete shut-down of the electrical system in case of an emergency. The shut-off switches should be

These fuel pump kill switches are a good idea to have on a car even if the rules do not require them. The switch shuts off the fuel pump if the motor runs out of oil pressure.

clearly marked to be visible from the outside of the car, so they are readily accessible to rescue workers in case of an emergency. Shut-off switches come in handy when working on the car's electrical system and to shut the whole system off when the car is in storage.

To charge your battery in between races get a small charger at the auto store. A two or three amp charger is inexpensive and one will bring the battery back up to par overnight.

The Starter

Starters on race motors are called upon to do some severe cranking. Race engines run hot and their compression is generally high, which puts a heavy load on the starter and the battery. Furthermore a race car might set for weeks or even months, which can cause some corrosion. A high performance starter such as the Tilton Super

The Tilton Super Starter used on our motor is the same as used on small block Fords. These starters are light and have great cranking power even when hot. Super Starters can be used with or without the solenoid.

When using the stock starter apply some silicone gel to the inspection plate and use a lock nut on the 10/32 bolt. Vibrations have been known to loosen the plate.

Starter would be a good investment. These starters are very powerful and durable. Tilton's Ford 2.3 unit is the same as for the Ford 351 V8. Specify your mounting face to ring gear distance when ordering a Super Starter so it can be spaced correctly.

If you run the stock starter, at least get a new one, or a good rebuilt such as the ones available from Ford dealers. The Ford rebuilt starters carry part number D6FR-11002-B. To prepare the starter, install a lock nut on the 10/32-inch bolt that holds the inspection plate around the starter and on the cable bolt.

The Ford starter solenoids work well on race cars, but take care to keep the solenoid away from engine heat, or you might have some cranking problems.

The Ignition System

The ignition system does not create any horsepower. The ignition only unleashes the power that is already present in the motor. An ignition system that breaks down at 7000 RPM limits the motor to that power range. If a good state- of-the-art electronic ignition system is used on the same motor and it makes power to 8000 RPM, it was not the ignition that created the power, the power was already in the motor.

On our car we used the Ford Motorsport distributor, part number M-12127-A231, the CD ignition module, part number M- 12199-A351 and the ignition wire set, part number M-12259- A221. The coil is from MSD (which also makes the Ford ignition units) and it is part number Blaster 2.

Ignition Installation

Electronic ignition is not as complicated to install as it

may appear. A few rules must be followed or the system will cause some weird problems.

1)The first and most important thing to remember is that these systems need perfectly good grounds. Clean all the terminals and mounting surfaces with some sandpaper and be certain the battery is well grounded to the frame of the car. 2)The distributor must be well grounded also. This is not a problem if you run solid mounts on the engine, but if you are running the stock mounts, run a good ground strap to the frame.

3)The next rule to follow is to run the heavy red wire from the ignition box to the battery. It is more convenient, though, to run this wire to the battery cable on the starter solenoid. The smaller red wire can be mounted to the on-off switch, but not to an accessory. Mounting the wire to the switch will ensure that the unit is getting all the voltage available.

4)If you must do some electrical welding on the car while the ignition box and the distributor are installed, take extra care to have a good ground on the welder and clamp the ground as near to the weld as possible. Never place the ground with the ignition module or the distributor between it and the weld. In this case the electricity would flow from the welding tip to the ground, passing through the distributor and the ignition module, which in itself will not cause a problem, but if the ground is poor, the electrical charge will arc through the distributor and the module. I fried a module and a distributor doing this. Incidentally, MSD has a very efficient and fast repair department!

5) Use the coil recommended for the ignition module you are using and do not use the high output coils with stock or breaker ignition. The heavy duty coil will burn the points in a few seconds. The ballast resistor supplied with the high output coil is for use with point ignitions. 6) To take maximum advantage of the increased spark output of the electronic ignition, set the plug gap wider than with a breaker ignition system. MSD recommends as much as .065-inch gap on the plugs, but MSD also warns that the wider the gap is, the greater are the chances of spark scatter. We had no problems running the 2.3 at .060- inch plug gap.

7)Install the module in a cool area, away from engine heat and vibration. We installed the box on the floor, in front of the driver's seat and we used the rubber mounts supplied by MSD. The coil should also be installed in a cool area, preferably with a cool air flow. 8) Use Spyro Wound or Helicore plug and coil wires only with the electronic ignitions.

The electronic ignition systems have often been maligned, but the real culprit is usually the operator or installer. Nevertheless, problems can still arise and the source must be traced. The module is usually blamed first, but the distributor, or coil, can also be at fault. Here is a simple test to pinpoint the problem if the coil is not firing: Take the coil wire about 1/2-inch out of the distributor cap and place it near a good ground. Next, disconnect the green and violet wires coming out of the module wiring loom. Switch on the ignition and tap the green and violet wires. If a spark appears at the coil wire, the distributor is defective. If you still do not get a spark, the coil or the module may be inoperative. Try a new coil. Just for the test, you could use the coil out of your passenger car. If you still do not get a spark, the module is probably defective, you have a bad ground, the battery is low or dead or the wiring is improperly connected. Often times the fuse link blows in the module when it experiences an over load such as experienced when a ground is loose or if welding with a bad ground has been done on the car. The module can be tested at some MSD dealers or it can be sent directly to MSD for repair.

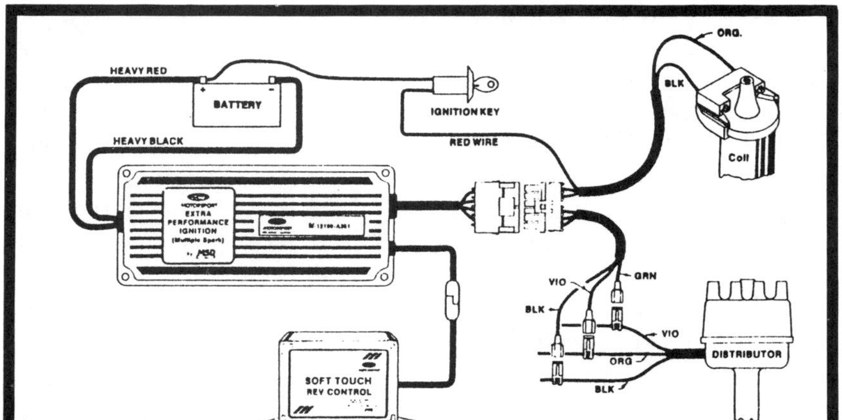

Wiring diagram for Ford SVO/Autotronics electronic ignition system.

Chapter 17
The Cooling System

The Radiator

We used a Modine radiator and oil cooler. These units are light, strong and they have excellent heat exchange ability. The radiator is model 748, Universal Downflow, while the cooler is model 1B-111. Both the cooler and the radiator are overkills, but it is easier to increase the coolant and oil temperatures, if needed, than it is to have to fight with an overheating engine.

Modine's aluminum radiators are vacuum brazed which improves hot strength tremendously. Silver solder has a relatively low melting point, and it is nowhere near the temperatures encountered in a cooling system, but as that temperature is approached, even in the 200 to 250-degree range, the joints start to weaken. Vibrations just add the finishing touches and the radiator starts to leak.

Modine recommends mounting both the radiator and the oil cooler at a 15 degree angle, top angled rearward. This angle of attack causes the incoming air to swirl around the fins, picking up more heat, before leaving the radiator. In addition the incoming air should be shrouded by a shield made of aluminum or plastic. Air follows the path of least resistance and if it is not channeled, it will simply flow around the radiator and cooler. What happens when a shroud is built around an air intake is a high pressure area is created in front of the radiators. If the high pressure is not bled-off behind the radiator or cooler, the air will not flow through and it will not be able to carry any heat away. Modine explains that radiators and coolers are heat exchangers, and they take the heat from the engine via the coolant and expose it to the air. The air is heated by the coolant and the coolant picks up the coolness from the air. If the air does not circulate, it will overheat and it will be unable to transfer any heat.

One of the key phrases in the explanation above is that the air must circulate. To circulate, the air must be able to enter the front of the radiator and it must also be able to bleed off the back. All this really means to the race car

Our Modine radiator and oil coolers fit nicely next to each other. The concern was to get the two coolers directly in the air stream and to leave an opening on the left to cool the carburetors and the coil. Note how the radiator and the cooler are mounted.

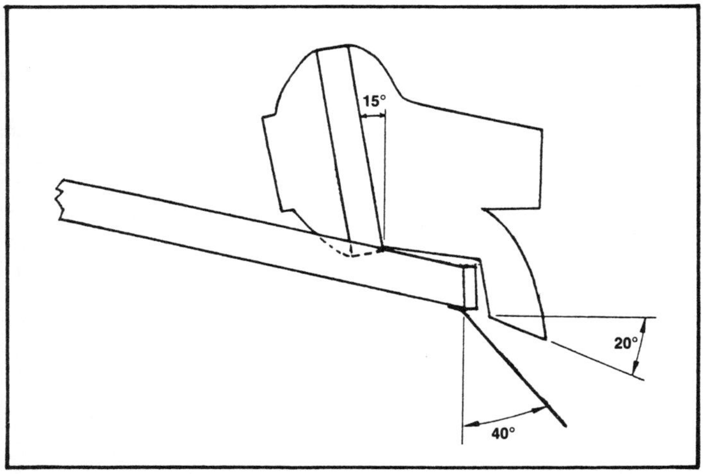

The proper method of mounting the radiator (15 degrees laid back) with a fan shroud behind it, and ducting which directs air into it from the grille and the front spoiler.

builder is that the air exit is just as, if not more, important than the entry. An oil cooler stuck under a fender with all the available air coming in, but with no place to go, will not cool!

The best place to locate the oil cooler on a Mini Stock is right in front of the engine, next to the radiator. The back end of tube chassis Minis is wide open and it acts as a big plenum for the air to escape.

The hot air behind the radiator should be channeled away from the carburetors and the engine electrical system. Build a shield to the left of the rear of the radiator to direct the hot air to the right side of the car, and run a separate vent to the carburetors and the coil. Do not worry about heating the oil if you run a dry sump, because the oil is hotter than the air going out of the radiator and cooler.

Pressure Capacity

Water boils at sea level at 212 degrees Fahrenheit. When pressure builds up in the cooling system, so does the boiling point. For each pound of pressure increase in the cooling system, the boiling point goes up by 3 1/2 degrees. So if you use a 22 PSI cap and the pressure rises to that mark, the boiling point will become 289 degrees F. The trick is not to build pressure by heating the coolant, but rather by pumping pressure into the system. This method is used on some Indy cars and the pressure is introduced in the system with a small pump. The idea behind this is to prevent hot spots. Hot spots are created in the head around the exhaust area because of the high temperatures created by the exhaust. The coolant might read a cool 200 or 220 on the gauge, but in the head the hot spots might get up to 300 or 325 degrees near the exhaust ports, causing the coolant to steam. Steam is not a liquid, it is a gas. Liquids have the ability to cool, but gases do not.

Use a Stant racing radiator cap, part number XL 2. These caps are available at most auto part stores and they must be special ordered. The capacity of these caps is 22 to 24 pounds.

Coolant

Run at least fifty percent coolant. This coolant should be a permanent-type ethylene glycol base. This anti-freeze solution raises the boiling point of the water and it has the added advantage of being a rust inhibitor and water pump lubricant. Anti-freeze also prevents foaming and cavitation.

Never use any stop leak in the cooling system. The stop leak gums up the tubes in the radiator and reduces its efficiency.

Mounting the Radiators

The radiator and cooler should be mounted in a semi-rigid cradle to prevent vibrations. Design the mount so it will bend if the car is hit, saving the radiator and cooler from damage.

Build a cradle under the radiator and cover it with rubber to absorb vibrations and shock. Remember that a radiator full of water is very heavy, so the mount should be strong enough to support the weight. On the top, use another cradle to hold the radiator down in place. The top cradle also must be padded. On our car we built a cradle for the top of the radiator that is hooked up on the back with a spring-loaded bolt. This way the radiator is not bolted solid, but spring mounted. The oil cooler and the radiator on our car are mounted together side-to-side on the same supports.

Fans

Stock fans should never be used on race cars. They are designed to work at very low RPM and they stall at the kind of engine speeds used on race cars. This stalling builds a wall behind the radiator and it actually reduces air flow. The other disadvantage these with fans is that they require a lot of power from the engine to turn them. Some passenger car fans draw as much as 20 horsepower. If you

ANTI-FREEZE CHART

Anti-freeze solution by volume (%)	Boiling point with 18# pressure cap
30	268°F.
40	270°F.
50	273°F.
60	278°F.
70	284°F.

The stock water outlet can be cut out and fitted with a water elbow made of tail pipe tubing. Preferably, use bent tubing rather than welded 90 degree fittings like this one.

use a fan, get a plastic blade fan and cut the blades down to reduce the drag.

Another way to reduce power losses and still get the cooling air flow is by installing an electric fan. These can be purchased from BAP/GEON auto parts stores. The fans come with thermostats that can be pre-set to your choice of temperature or they can be switched on manually.

Coolant Restriction

A water pump, at racing speeds, turns too fast to allow the coolant enough time in the radiator to cool itself. The solution is to slow down the pump using a larger water pump pulley or a smaller crankshaft pulley. These crankshaft pulleys are manufactured by Esslinger Engineering and they slow down the water pump by 30 percent. These pulleys offer the added advantage of being graduated with timing marks.

The coolant should also be restricted at the thermostat housing by using a 5/8-inch I.D. restrictor plate. This will allow the coolant to spend a little extra time in the head to pick up the heat. Thermostats should always be removed as they can stick and a thermostat may over restrict the flow of coolant.

Most engines put out their best power at around 200 to 220 degrees of water temperature.

Hoses

As much as possible use the Gates Green Stripe radiator hose as it has great abrasion, heat and pressure resistance. This hose can be purchased in three foot lengths from truck repair shops or parts houses. Unfortunately this hose only comes in straight lengths. To make your bends, use bent tail pipe tubing or angled radiator hose. If you need to go from 1 3/4 inches out of the radiator to 1 1/2 inches to the pump, you can use a brass plumbing reducer from any house plumbing supply store. For the lower hose you must use a hose reinforced with a spring on the inside. The spring gives the hose the rigidity it needs to accept the negative pressure generated by the suction of the pump.

This hose would fly off the outlet if the pressure should blow. Use a heavy overflow line and clamp it.

Use good quality clamps such as the Ideal stainless steel clamps, also available at truck repair shops. Another good clamp is the Marman clamp sold by Baker Precision.

The Catch Can

The coolant recovery can is not only important to have for the proper functioning of the system, but it is a rule with most racing associations. The recovery can serves two purposes: 1) If the engine overheats and some coolant is lost, the can will catch the coolant that escaped from the radiator, and when the engine cools down, the radiator will form a vacuum which will draw the coolant back into the system. 2)The catch can keeps the coolant off the track if the engine overheats.

The can should still be vented in a careful fashion because if it should fill up, it will still leak. Vent the can to the atmosphere through a hose of at least 1/4-inch diameter, as too small a hose could build up a restriction from the escaping steam and water and it could actually blow off. Aim the overflow hose toward the windshield so you can see it if it should blow. Run the hose all the way to the right side of the windshield, so if any of the hot water or steam seeps through the body, it will not burn the driver.

Chapter 18

Junk Yard Parts

Before rushing off to the wrecking yard, consider purchasing a used Pinto or Mustang. We picked up a 1974 Pinto for $200, and the thing even ran! The car gave us all the parts we needed for the engine, trans (if you choose to run the four speed), the front end and the front brakes (if you cannot afford the racing units). Odds and ends off the car also came in handy, such as the rear view mirror, the gear shift knob and some of the metric fasteners. The rest of the car was traded for some welding, so the $200 went a long way.

Buying a complete car has another advantage. You can see if it has been hit severely in its past, and the car can be driven to make sure the block or head are not cracked. From a wrecking yard you never really know what you are getting.

Try to get your parts from a yard that allows you to remove the parts off the car yourself. Inspect the car for damage and never choose parts, even engines or transmissions, from badly wrecked cars. Also never buy parts off cars that have been driven in the snow. Salt used on icy roads causes severe corrosion which in turn weakens parts, not to mention the fact that bolts will probably be frozen-up solid. And, even if they will budge, they will probably break in their hole.

Conduct a complete inspection of the parts before buying them. It will save you a lot of trouble and hassle in the long run. Be sure to find out in advance what the yard will guarantee. Try especially to get a written guarantee on major components against cracks and major problems. If any of the parts are damaged and unusable, return them immediately for an exchange or refund.

The parts you will need from a bone yard are: engine, transmission (see transmission section in this book for additional suggestions), steering rack with tie rods and ends, spindles, hubs, discs and calipers. If you have a professional engine builder do your motor, you can do without that purchase.

If you run a 2300 CC motor, you will certainly want to use the Ford 5 speed, and if you run the aftermarket brakes, discs, hubs and spindles, you will not need those parts either. The stock Pinto rear end is marginal but behind a mild 2000cc motor, it should serve the budget racer rather well.

Once the parts have been purchased, they should be steam cleaned and disassembled for inspection, and the spindles should also be "Magnafluxed". The engine should be dismantled, cleaned and sent out along with the head to test for cracks. The calipers should be taken apart and inspected. Again, cars that have been driven in the snow will probably have damaged calipers. Inspect the bores for pits and gouges. Look at the wheel bearings and races. The bearings may be retained if they look brand new or if you are on a real tight budget and you plan to just test the car on the old bearings.

The Ford 9-inch Rear End

These rear end assemblies have been the work horse of

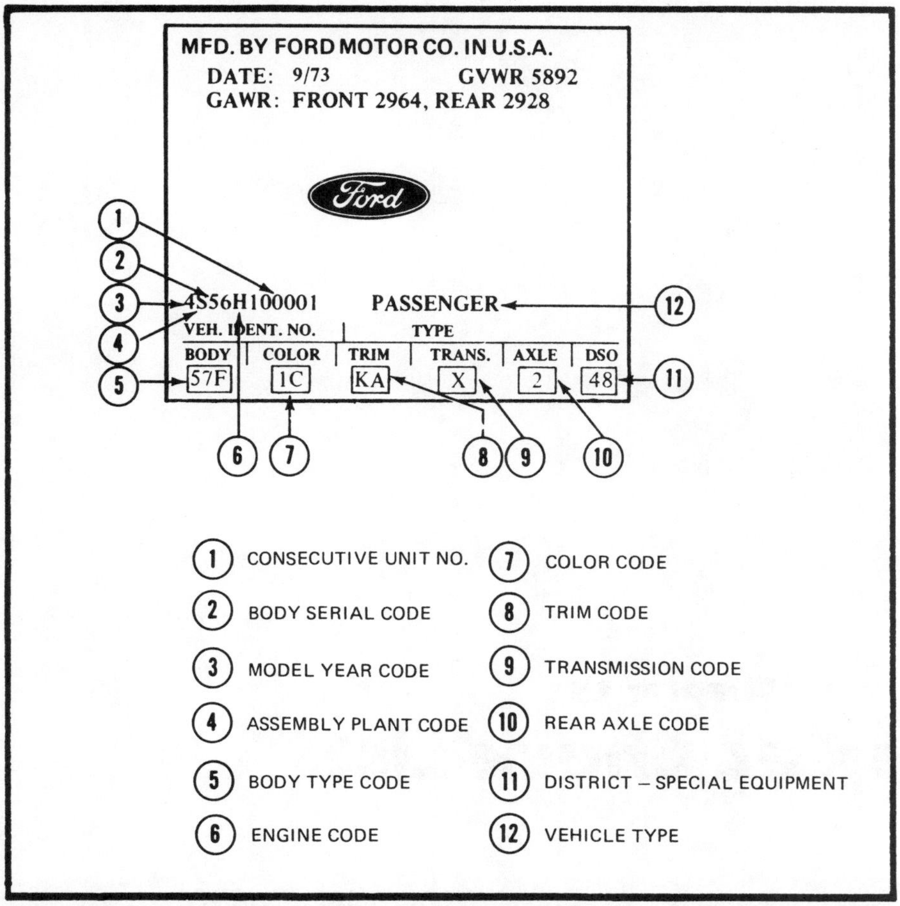

Ford product vehicle identification numbers and interpretation.

stock car racing for many years for racers from the Saturday nighter to the all-out pro. The "nine inch" can be purchased used from wrecking yards, it can be found new in the Ford Motorsport catalog (M-4001), along with about thirty different ratios, or the racer can purchase a Ford nine inch, all set-up, from Speedway Engineering. This rear end is widely available, easy to work on, very strong, and the range of ratios available is very wide.

The easiest way to spot the 9-inch rear end is, first of all, by its removable carrier. However there are four different removable carrier Ford rear ends: 8-inch, 8 3/4- inch, 9-inch and 9 3/8-inch ring gears. Of these four, two can be immediately identified as not being the 9-inch.

The 8-incher is obviously small, and it is found in lighter weight vehicles. The 9 3/8- inch carrier has a pinion snubber cast on the housing above the companion flange as an integral part of the carrier, whereas the 9- inch housing uses a bolt-on snubber.

When the choice is narrowed down to a suspected 9-inch ring gear rear end, the carrier can be pulled and the ring gear measured. There is a stamped metal product identification tag attached by a carrier housing bolt on all 1963 and later Ford axle housings. The code stamped on that tag will help you determine the ring gear diameter, application and ratio.

A Ford parts man can make the interpretation for you, or the wrecking yard should have a Hollander Interchange Manual which identifies all the codes.

Once you have everything narrowed down to the 9-inch ring gear carrier, there are some differences to sort out. First of all, there are two different size pinion carrier bearings: 2 57/64-inches O.D., and 3 1/16-inches O.D. The larger bearing is preferred for racing use. Secondly there have been four different sizes of pinion gear roller bearings. All work equally well for any application, but if you have several rear ends, you should try to keep all the bearings the same size so you will not have to stock all the various bearings.

Several carrier housings can be used with the 9-inch rear end. The most desirable is the nodular cast iron case which is easily identified with an "N" cast on it. There are also high nodular iron cases which are acceptable and they have the cross hatch ribs cast in the housing, but not the characteristic "N".

There are also several aftermarket manufacturers that produce 9-inch carrier housings, both in high nodular cast iron and alloy aluminum (high quality, lower unsprung weight). These can be obtained from Mark Williams Enterprises and several other manufacturers.

Chapter 19

Race Car Check List

We have provided a checklist here as an aid to the racer in caring for his car. The important point is that proper preparation, organization and attention to details are most times the key to winning races — not the factor of brute horsepower or driver skill. It has been said before, and we will say it again — a car has to finish a race to win it.

We all can recount stories we know of about drivers who have lost races because they have failed to check oil levels, brake fluid levels, wheel bearing conditions, critical hardware tightness, etc. It all happens through carelessness and/or forgetfuness. Our check list is intended as one step towards eliminating those problems.

The list secondly is a check which should be followed through before the car goes racing each week. Sure, it can be said that some of the items are elementary and may insult the intelligence. But we can answer with the stories of well-known professional racing teams which have sent their cars into competition without fueling the car, or having any lube in the differential. A rigidly adhered to check list probably would have let those cars finish a race.

You might want to make up lists for yourself which you can tape to your windshield each week to be sure everything is checked.

WEEKLY MAINTENANCE

A. Cockpit

1. Are gauges operating properly?
2. Belt and harness condition and mounting
3. Seat mounting
4. Fire extinguisher — charge condition and mount hardware
5. Interior wiring
6. Cracks developing in roll cage tubing or welds
7. Tachometer cable tight
8. Fuel pressure gauge line tight
9. Cockpit clean, including glass or Lexan
10. Mirror mount solid

B. ENGINE

1. Oil level
2. Oil condition
3. Water level
4. Sparkplugs — gap and wear
5. Plug wires
6. Distributor cap — inside pole wear, cracking, arc tracks
7. Ignition high tension leads
8. Points
9. Valve lash
10. Pulley belts
11. Crank dampner
12. All external fuel and oil lines

13. Water hoses
14. Valve covers
15. Radiator core
16. Oil cooler — core and connections
17. Headers
18. Fuel filter
19. Fuel line from cell
20. Oil filter
21. Air filter
22. Throttle linkage
23. Throttle return springs
24. All carburetor bolts and screws — tightness. Also check tightness of boosters.
25. Carburetor fuel pressure
26. Cylinder head bolt torque
27. Compression check
28. Leak check
29. Timing
30. Fan bolts tight
31. Fan belt tension

C. SUSPENSION

1. Brake shoes or disc pads wear
2. Brake drum condition or rotor condition
3. Disc brakes
   a. Caliper bolts tight
   b. Hat bolts tight
4. Wheel cylinder condition
5. Wheel bearings condition, lube and preload
6. Brake system bleed
7. Master cylinder fluid level
8. Shock absorbers condition and mounting condition
9. Trailing arm bushings and hardware
10. Panhard bar — hardware and rod end condition
11. Panhard bar brackets — cracking
12. Crossmember cracking
13. Spring seats
14. Wheel studs condition
15. Upper and lower A-arm bushings condition and lube
16. Ball joints condition
17. Springs — true rate
18. Anti-roll bar
   a. Mounting bushings condition
   b. Spherical bearing links condition
19. Upper and lower A-arms attaching bolts
20. Steering linkage
21. Tie rod ends condition
22. Wheels condition
   a. Cracks
   b. Valve stems condition

D. ELECTRICAL

1. Battery charge condition
2. Battery connections
3. Battery securing hardware

E. DRIVE LINE

1. Clutch condition
2. Clutch free play
3. Clutch linkage condition
4. Clutch slave cylinder leakage
5. Clutch master cylinder fluid level
6. U-joints condition
7. Transmission leakage
8. Transmission case cranks
9. Transmission crossmember condition
10. Rear end leakage
11. Shifter linkage
12. Throw out bearing
13. Rear end lube level and condition
14. Transmission lube level and condition

F. CHASSIS GENERAL

1. Ballast secure
2. Bumper attachments secure
3. Fuel cell secure
4. Fuel line condition
5. Exhaust pipe brackets condition
6. Brake lines condition
7. Sheetmetal condition
8. Appearance
9. Hood pins and hood pin cables secure

G. CHASSIS ALIGNMENT

1. Weight at each corner
2. Chassis height at each corner
3. Caster
4. Camber
5. Toe-out

H. AT THE TRACK

1. Check lug nut tightness — most important
2. Spark plugs
3. Timing
4. Tire pressures
5. Fuel level
6. Oil level
7. Tire temperatures after warm-up

Suppliers Directory

Aeroquip
300 South East Ave.
Jackson, MI 49203
High performance hoses and fittings.

ARP
8565 Canoga Ave.
Canoga Park, CA 91304
(818) 341-4488
Heavy duty fasteners for the engine and chassis

ATL
Spear Rd.Industrial Park
Ramsey, NJ 07446
(201) 825-1400
Safety fuel cells and filling equipment

Autosigns Unlimited
44659 Yucca
Lancaster, CA 93534
(805) 942-0621
Custom decals of any size, shape and color

BAP/GEON
3025 E.Victoria St.
Rancho Dominguez, CA 90221 (Stores nationwide)
Small car and four cylinder auto parts stores.

Bart Wheels
P.O. Box 30
Conyers, GA 30207
(404) 922-1983
Racing steel wheels

BRC
7091 Belgrave St.
Garden Grove, CA 92641
(714) 893-0595
Custom pistons and crankshafts

Baker Precision
2865 Gundry
Long Beach, CA 90806
(213) 595-7208
Rod ends, braided lines, fittings, steering u-joints, System 1 oil filters, clamps, lubricants, coatings

Berthold Transmissions
395 St.Regis North
St. Constant, Quebec
Canada JOL1X0
(514) 638-2960
Racing two speed transmissions for oval track racing

Berge Mathews
5826 S Broadway
St. Louis, MO 63111
(341) 481-3836,
(714) 656-3686
Manual water thermostat

Cal Aero Supply
13840 Paramount Blvd.
Paramount, CA 90723
(213) 634-0383
Surplus hardware and fasteners

Cal Smith
(818) 357-2951
Race car bodywork and custom painting

Carrera Shocks
5412 Peachtree Rd.
Atlanta, GA 30341
(404) 451-8811
Shocks, springs and coil-overs

Champion Spark Plug Co.
P.O. Box 321
Toledo, OH 43661
Spark plugs

Canton Custom Oil Pans
9 Tipping Dr.
Branford, CT 06405
(203) 481-8324
Wet and dry sump oil pans, standard and custom

Colley Ford
655 S. Grand Ave.
Glendora, CA 91740
(818) 963-4131
Largest Ford Motorsport dealer in the US

Crane Cams
Box 160
Hallandale, FL 33009
(800) 328-1006
Camshafts and valvetrain components

Cyclone Headers
19914 S.Via Baron
Compton, CA 90220
(213) 638-7883
Exhaust headers

Cylinder Head Abrasives
2120 Blumenfeld Dr.
Sacramento, CA 95815
(916) 922-7652
Abrasives for grinding, polishing or sanding

Dana Corp
P.O. Box 321
Toledo, OH 43691
(419) 866-2630
Steel and aluminum driveshafts and u-joints

Del West
9440 Irondale Ave.
Chatsworth, CA 91311
Titanium valves

Digatron
North 6463 Pery St.
Spokane, WA 99207
(509) 487-2715
Digital tire pyrometers and exhaust gas temperature gauges

Earl's Supply Co.
14611 Hawthorne Blvd.
Lawndale, CA 90260
(213) 679-1438
AN fittings and hose

Ebeling Engineering
2503 Graydon Ave.
Monrovia, CA 91016
(818) 359-3807
Mini Stock and Modified chassis and complete race cars. Suspension components and race car fabrication.

Edmiston Engineering
107 N. Boulder
Tulsa, OK 74103
Tire pressure relief valves

Electramotive
140 Center St.
El Segundo, CA 90245
(213) 322-3222
Aircraft wiring, switches, circuit breakers and connectors

Esslinger Engineering
712 Montecito Dr.
San Gabriel, CA 91776
(818) 289-3073
Racing Ford four cylinder engines for all applications. Also parts and components for the Ford four cylinder engines, dyno work, head porting, Weber carburetors, manifolds, oil pumps, dry sumps, turbos, headers, flywheels.

Fast-Fab
920 S.Spring St.
Klamath Falls, OR 97601
(503) 884-2565
Inexpensive race car scales

Fuel Bladders Inc.
4231 Spencer St.
Torrance, CA 90503
(213) 542-3835
Fuel cells and components

Fel Pro
7450 N.McCormick Blvd.
Skokie, Il, 60076
High performance gaskets

Flexi Flyer
P.O. Box 34641
Phoenix, AZ 85067
(800) 528-2059
Race car scales

Ford Dealers Association (Contingency)
17000 Southfield Rd.
Allen Park, MI 48101
Contingency awards for Ford drivers

Ford SVO/Motorsports
17000 Southfield Rd.
Allen Park, MI 48101
(313) 337-1356
Ford competition parts for the engine, body and drivetrain. Also gauges, clothing and accessories.

Gates
999 S. Broadway
Denver, CO 80217
(303) 744-1911
High performance belts and water hoses

Genesis
14594 Cerces
Fontana, CA 97355
(714) 350-1705
Fiberglass racing seats

Goodyear
Carrol Shelby Enterprises
3200 E.69th St.
Long Beach, CA 90805
(213) 538-2914
Goodyear Racing tires

Grant Products
700 Allen Ave.
Glendale, CA 91201
(213) 849-3171
Custom and racing steering wheels

JFZ Disc Brakes
9761 Variel Ave.
Chatsworth, CA 91311
(818) 882-4717
Race car brakes and components

K&N Engineering
561 Iowa Ave.
Riverside, CA 92502
(714) 682-8813
Racing air filters and breathers

Leon Ruther Ent.
(818) 366-5693, 841-9190
Stainless steel dry sump pump manifolds and general stainless steel fabrication

Loctite Corp.
4450 Cranwood St.
Cleveland, OH 44128
(800) 243-8160
Gasket sealers, shake-proof epoxies and glues

Manley
13 Race St.
Bloomfield, NJ 07003
(800) 526-1362
Competition rocker arms for the Ford OHC motors

Mecca
Route 41
Sharon, CT 06069
(203) 364-5130
Pre-oiling sytems and oil pressurizing systems

Mc Cord
2850 W. Grand Blvd.
Detroit, MI 48202
High performance gaskets

Mc Creary
McCoy Racing Products
553 S.7th St.
Modesto, CA 95350
(209) 521-5656
Mc Creary racing tires, Carrera shocks and springs and race car hardware

Modine
1500 De Koven Ave.
Racine, WI 53401
(414) 636-1222
Competition and high performance radiators, oil coolers and rear end and transmission coolers

Moroso
P.O. Box 1470
Guilford, CT 06437
(203) 453-6571
In line screen filters, heat shields, rear end filler cans, adjustable oil pressure warning light, battery disconnect switch, quick release steering wheel hub, roll cage gussets and general race car hardware.

MSD Electronic Ignitions
Autotronics Corp.
6908 Commerce
El Paso, TX 79915
(915) 772-7431
Electronic ignition systems

Oberg
12414 Hwy.99 #55
Everett, WA 98204
(206) 778-1227
Screen type oil and fuel filters

Patterson
7049 Valjean Ave.
Van Nuys, CA 91406
(818) 781-1421
Dry sump oil tanks, heaters and hardware

Peterson Automotive Products
4485 S.Broadway
Englewood, CO 80110
(303) 781-7290
Dry sump pumps, accessories and fuel pressure regulators

Phoenix Fire Systems
1609 S.Central Ave., Bay 1,
Kent, WA 98032
(206) 852-5035
Race car automatic fire systems

Pitts Performance
7141 Woodley Ave.
Van Nuys, CA 91406
(818) 780-2184
Digatron digital tire pyrometers and digital exhaust temperature gauges.

PRO Fuels
1463 E. 223rd St.
Carson, CA 90745
Quality high octane racing fuels

Pyrotect
2900 E.42nd St.
Minneapolis, MN 55406
(612) 721-1646
Driver's fire and safety equipment

Redline Engineering
303 W.Artesia
Compton, CA 90224
(213) 637-7774
Weber carburetors, manuals, accessories and manifolds

Roger Kraus Racing
2896 Grove Way
Castro Valley, CA 94546
(415) 886-4636
Race car tires, wheel balancer and tire mounting tool

Shaver Specialties
20608 Earl St.
Torrance, CA 90503
(213) 370-6941
Dry sump pump drives, sprint car race motors

Simpson Safety Equipment
22412 S.Normandie Ave.
Torrance, CA 90502
(213) 320-2204
All types of safety equipment

So-Cal Racing Fuel
8504 E.Compton Blvd.
Paramount, CA 90723
(213) 408-0440
Racing fuels and additives

Speedway Engineering
13040 Bradley Ave.
Sylmar, CA 91342
(818) 362-5865
Mini quick change rear ends, floater rear ends, steering racks, corner weight checkers, gears, chassis, hardware

Stagger Master
13425 S.E. 27th Pl.
Bellevue, WA 98005
(206) 746-1154
Stagger measuring device

Stick Only
9900 Glenoaks Blvd.
Sun Valley, CA 91352
(818) 768-4747
Standard transmission modifications for racing

Stock Car Fiberglass
P.O. Box 4504
Columbus, GA 31904
(404) 323-1101
Stock car fiberglass bodies

Streaker
5636 Shull St.
Bell Gardens, CA 90201
(213) 806-3347
Split rim aluminum racing wheels

Supertrapp
1222 Research Park Drive
Davis, CA 95616
(916) 756-5069
Racing mufflers

System 1
1822 E.Main, Suite A
Visalia, CA 93291
(800) 231-9137 In Calif.
(800) 554-3533 (U.S.)
Inspectable oil filters for racing and street

Tex Racing Ent.
PO Box 126
Ether, NC 27247
(919) 428-9522
Transmission racing preparation and complete racing cars

Tilton Engineering
P.O. Box 1787
Buellton, CA 93427
(805) 688-2353
Racing pedal assemblies, flywheels, master cylinders, brake remote reservoirs, hydraulic clutch systems, heavy duty starters, racing clutches, brake bias adjusters

Total Seal
2225 W. Mountain View, #16
Phoenix, AZ 85021
(603) 436-3887
Competition gapless piston rings

TRW Automotive
8001 E. Pleasant Valley Rd.
Cleveland, OH 44131
Competition pistons, valves, rocker arm studs, stock Ford steering racks, fuel pumps

Tubes Inc.
6489 Corvette St.
Los Angeles, CA 90040
(213) 726-9793
Round and square tubing for race car frames and chassis

Usinage Berthold Engineering
395 St.Regis North
St. Constant, Quebec
Canada JOL1XO
(514) 638-2960
Two speed racing transmissions

Vic Irvan Ent.
80 Lowe Ave.
Concord, NC 28025
(704) 788-2554
Bump steer gauge, pit equipment, Cleco fasteners, rivets, air riveter, safety wire pliers, rod ends, sheet metal brakes

Wayne Mitchell Engineering
11404 Sorrento Valley Rd.
San Diego, CA 92121
(619) 452-0961
Camber/caster gauges

Notes

Notes

Notes

Notes

STEVE SMITH AUTOSPORTS

PUBLICATIONS

The Complete Karting Guide

The complete guide to kart racing, for every class from beginner to enduro. Contains: Buying a kart and equipment • Setting up a kart — tires, weight distribution, tire readings, plug reading, exhaust tuning, gearing, aerodynamics • Engine care • Competition tips • And more. **#S140 . . . $9.95**

Racing The Yamaha KT 100-S Engine

Hi-perf. building and blueprinting one of the most popular two stroke engines. Includes: Two cycle engine basics • Blueprinting and tuning the entire engine • Tips for ultimate power • Building for reliability • Complete engine machining • Ignition system • Carburetor tips • Using alcohol • Expansion chamber and exhaust tuning • Maintenance. Full of photos. **#S151 ..... $9.95**

Complete Stock Car Chassis Guide

A more advanced stock car handling book. Covers: Lateral acceleration and weight transfer • Finding the CGH • Roll centers • Roll couple • Using anti-roll bars • Racing tire basics and selection • Shock absorbers selection • Aerodynamics basics • 2-point rear suspension. **#S102 . . . $6.75**

Bldg A Race Car Picture By Picture

A collection of detailed photos taken underneath, inside and under the hood of some of America's most successful stock cars. Shows the elements that make a truly winning car: Electrical and plumbing detail, construction details, component placement and much more. **#S110 . . . $6.95**

Dirt Track Chassis Blueprint

Includes all the latest technology to get you competitive on dirt — offset design, left side bias, rear bias. Includes full roll cage details. A rigid, yet lightweight design. Plans include details for a fifth coil-over rear suspension, or leaf spring rear. **#B500 . . . $17.95**

Bob Bondurant High Perf Driving

A fascinating book that teaches the necessary skills for high-speed driving at minimum risk. Bondurant translates track-proven techniques into 12 chapters about getting the most out of yourself and your car and controlling it! Learn to do it from the master. Over 100 illus. 144 pages. **#MB35 . . . $11.95**

Racer's Complete Reference Guide

The "Yellow Pages" of the high performance world. Where to get: Hardware, chassis and engine parts, running gear, etc. Where to find: Fabricators, engine builders, safety gear, racing associations, driving schools, etc. A unique and valuable reference source for racers. 430 illus. **#S108 . . . $8.95**

How To Run A Successful Racing Business

How to operate a racing-oriented business successfully. How to: Start a business • Put more profit in a business • Cost controls • Proper advertising • Dealing with employees, taxes, and accountants • And much more. The secrets of business fundamentals and how to make them work for you. For existing businesses too. **#S143 . . . $9.95**

Buick Free Spirit Power Manual

Complete information on the performance engine of the future — the Buick V6. Blocks, crankshafts, rods, valve gear, intake and exhaust systems, ignition, susupension, brakes and body modifications. **All new** revised edition. Complete list of performance hardware. Over 200 photos and drawings. **#S123 ..... $9.95**

The Stock Car Racing Chassis

The basic stock car handling "get acquainted" book. Covers: Understeer and oversteer • The ideal chassis • Spring rates and principles • Wedge • Camber/caster/toe • Sorting a chassis with tire temperatures • Building a Chevelle chassis. **#S101 . . . $6.75**

Building A Street Stock Step-By-Step

Step-by-step info from buying the car and roll cage kit to mounting the cage, stripping the car, prep tips, front and rear suspension modifications, adjusting the handling within street stock rules, at-the-track setup, and mounting seats, gauges, linkages, etc. A complete guide for the entry level street stock class. **#S144 . . . $9.95**

Race Car Graphics

Takes the "magic" out of race car coloring, numbering, painting, striping and graphic design by showing how you can do it. Make your car and operation look absolutely professional without spending a lot of time and money. Includes 8 pages of full-color photos. **#S118 . . . $7.95**

Racing The Small Block Chevy

A totally up-to-date high performance guide to the small block Chevy. Covers: High performance at a reasonable cost • Component blueprinting • Revealing cam, carb head porting and ignition tips from many of racing's best known engine builders • 330 illustrations. **#S112 . . . $9.95**

Racer's Guide To Fabricating Shop Equipment

Would you believe an engine stand, hydraulic press, engine hoist, sheet metal brake and motorized flame cutter all for under $400? It's true. Step-by-step instructions and concise photos and drawings show you how to easily construct them yourself. **#S145 ...... $9.95**

Stock Car Driving Techniques

Written with authority by GN driver **Benny Parsons.** Covers: Basic competition techniques • Getting started in racing • Improving your lap times • Defensive driving tactics • Every aspect of competition driving. **#S104 . . . $7.95**

Advanced Race Car Suspension

The latest tech information about race car chassis design, set-up and development. Weight transfer, suspension and steering geometry, calculating spring and shock rates, vehicle balance, chassis rigidity, and much more. **#S105... $9.95 Work Book for above book . . . #WB5 . . . $5.95**

Bldg The Chevy Sprint Car Engine

Includes complete info on the Donovan and Milodon aluminum blocks, all aluminum heads, fuel injection systems and set-up, burning alcohol, magnetos, plus complete chapters on blueprinting all components. **#S130 . . . $9.95**

Sprint Car Technology

A complete book on buying, building and racing a sprint car, midget or super modified. Covers: Torsion bars • Chassis structure • Straight front axle suspension • Live rear axles • DeDion rear • Birdcages • Lateral locating linkages • Steering • Bump steer • Shocks • Dirt and asphalt set-ups • Tire selection • And more. Photo packed. **#S125 . . . $11.95**

Racing Engine Preparation

By Waddell Wilson, one of the best engine builders in NASCAR. He takes you through all the steps of engine building as carefully as he builds one. Explains: Cam timing and selection • Head porting • Carb mods • Prepping rods, pistons and blocks • H.P. ignition and lube systems. "The most complete engine book I've ever seen," says **A.J. Foyt.** Uses small block Chevy, Ford and MoPar. **#S106 . . . $9.95**

Building The Hobby Stock Car

For the low-buck racer with limited facilities. How to build a car with a minimum of "store bought" parts or expensive machine work. Covers: Car choice • Cage construction • Unitized vs. full frame chassis • Engine, cooling and electrics • Transmissions and rear ends • Suspension (parts choice, building and sorting) • Driving • Preparation, and more. **#S126......$9.95**

The Racer's Tax Guide

How to save BIG money on your racing activities. This book tells you how to LEGALLY subtract your racing costs from your income tax bill by running your operation like a business and following 3 simple steps. An alternative form of funding your racing. Includes a concise, up-to-the-minute report on how the new tax laws and changes help you save even MORE money. Read this book **now** and start saving! It works! **#S116... $9.95**

Race Car Fabrication & Preparation

Includes thorough discussions of: Chassis and suspension fabrication and design • Preparing a trans • Setting up a rear end • Electrical system • Building the roll cage • Cutting costs and beating the economy stock rules • Welding • Clutches • Wrecking yard parts • Plumbing • Driveshafts • Wheels and tires. Plus much, much more. **#S114 . . . $10.95**

Coil-Over Suspension Booklet

An 8-page booklet showing how to design a proper chassis to use coil-over suspension units. Full of photos and drawings. **#BC201 . . . $5.95**

Dirt Track Chassis Technology

Covers EVERY facet of dirt track racing and stock car chassis set-up. Includes all the latest hot topics — left side bias, aerodynamics, leaf vs. coil suspension, and details on the 5th coil-over suspension. This massive new book is a MUST! **#S133 . . . $12.95**

Stock Car Chassis Technology

A brand-new suspension book by **Steve Smith!** The latest in race car suspension tips, all the newest ideas in stock car chassis set-up. Includes: Chassis set-up principles • Procedure for set-up at the shop, at the track • Actual case studies • Front suspension set-up and adjustment • Rear suspension — all the new torque absorbing systems, torque arms, stagger, locating devices • Repairing a crashed car • The newest on aerodynamics for the short tracks • And much, much more. For paved and dirt track applications. **#S139 . . . $12.95**

The Doctor's Guide To Automotive Ignition

The most complete book on high performance and automotive ignitions ever written. All about conventional and electronic ignitions. Includes: Hi performance from factory electronic ignitions • Improving engine and ignition performance • Creating custom tailored systems • Complete troubleshooting guide • Tech tips for ultimate ignition • and much more. **#J153 ... $12.95**

Trailers — How To Buy & Evaluate

A complete guide to: Buying a used or new trailer • Designing and engineering a trailer for your use • Trailer components, tongues, couplers, the right hitch • Selecting the right tires • Creating towability • Eliminating sway • Trailer brakes, suspension, stability. Well illustrated. **#S150 ..... $9.95**

Building a Mini Stock — Step By Step

From fabricating the tubular frame to installing the fiberglass body, this book covers it all. Plus a complete high performance build-up of the Ford 2000/2300cc 4 cylinder engine. Suspension, running gear, and at-the-track adjustment. Everything is here. **#S152 ..... $10.95**

So You Want To Go Racing?

Finally, a book written for the enthusiast considering his first start in racing. Gives a complete understanding of: Setting up a shop • Buying the right tools and equipment • Organizing and gathering a crew • Getting the most out of a limited budget • Being resourceful • Shop setup • Budget checklists • And more. This book is like taking an apprenticeship to a 20-year racing veteran. **#S142 . . . $9.95**

Ordering Information

- Call us and charge your order
- In Canada add $3.00 to all orders
- For 1st Class add $3.00 per book up to $9.00
- California residents add 6% sales tax
- Allow 3-4 weeks for 4th class delivery
- Checks over $30 must clear first
- Remit in U.S. funds

ORDER HOTLINE (714) 639-7681

Steve Smith Autosports/P.O. Box 11631/Santa Ana, CA 92711